Porous Fluids - Advances in Fluid Flow and Transport Phenomena in Porous Media

Edited by Vallampati Ramachandra Prasad

Published in London, United Kingdom

IntechOpen

Supporting open minds since 2005

Porous Fluids - Advances in Fluid Flow and Transport Phenomena in Porous Media
http://dx.doi.org/10.5772/intechopen.91125
Edited by Vallampati Ramachandra Prasad

Contributors
Paul Papatzacos, Twinkle R. Singh, Anindityo Patmonoaji, Tetsuya Suekane, Yingxue Hu, Zhang Chunwei, Kieu Hiep Le, Kazuhisa Yuki, Mustapha Sadouki

First published in London, United Kingdom, 2021 by IntechOpen
IntechOpen is the global imprint of INTECHOPEN LIMITED, registered in England and Wales, registration number: 11086078, 5 Princes Gate Court, London, SW7 2QJ, United Kingdom
Printed in Croatia

British Library Cataloguing-in-Publication Data
A catalogue record for this book is available from the British Library

Additional hard and PDF copies can be obtained from orders@intechopen.com

Porous Fluids - Advances in Fluid Flow and Transport Phenomena in Porous Media
Edited by Vallampati Ramachandra Prasad
p. cm.
Print ISBN 978-1-83962-711-8
Online ISBN 978-1-83962-712-5
eBook (PDF) ISBN 978-1-83962-722-4

For EU product safety concerns:
IN TECH d.o.o., Prolaz Marije Krucifikse Kozulić 3, 51000 Rijeka, Croatia, info@intechopen.com or visit our website at intechopen.com.

Meet the editor

Over the past decade, Prof. V. Ramachandra Prasad has worked in the areas of multi-physical fluids dynamics and numerical simulation. His work has investigated complex phenomena of interest in mechanical engineering, applied mathematics, chemical engineering, and material processing. He has made fundamental contributions in magnetic fluid dynamics, thermal radiation heat transfer, fluid modeling, micropolar non-Newtonian hydrodynamics, magnetic induction flows, viscoelastic liquid flows, porous media, and boundary flows. He has published more than 150 journal articles in these areas and has made numerous conference presentations. He has published six books on radiation heat transfer, thermo-diffusion and diffusion-thermo effects boundary layers' flows, and viscoelastic flows. His more recent work has involved non-Newtonian fluid dynamics, fluid-particle suspensions, and lid-driven cavities. He has used the numerical methods of finite difference, finite element, network simulation, and finite volume codes to solve problems in his areas of research in collaboration with US, Indian, Spanish, British, and Taiwanese researchers. He is a professor of Mathematics at Vellore Institute of Technology, Vellore, India.

Contents

Preface XI

Section 1
Multiscale Modelling in Porous Media 1

Chapter 1 3
Porous Flow with Diffuse Interfaces
by Paul Papatzacos

Chapter 2 25
Multiscale Modeling of Non-Isothermal Fluid Transport Involved in Drying Process of Porous Media
by Kieu Hiep Le

Section 2
Mathematical Models in Porous Media and Solutions 43

Chapter 3 45
Study on Approximate Analytical Method with Its Application Arising in Fluid Flow
by Twinkle R. Singh

Section 3
Experimental Challenges in Geological Applications of Porous Media 59

Chapter 4 61
Dissolution Mass Transfer of Trapped Phase in Porous Media
by Anindityo Patmonoaji, Yingxue Hu, Chunwei Zhang and Tetsuya Suekane

Section 4
Heat and Mass Transfer in Porous Media 83

Chapter 5 85
Heat Transfer Enhancement Using Unidirectional Porous Media under High Heat Flux Conditions
by Kazuhisa Yuki

Section 5
Modelling of Porous Media and Systems **105**

Chapter 6 **107**
Inverse Measurement of the Thickness and Flow Resistivity of Porous Materials via Reflected Low Frequency Waves-Frequency Approach
by Mustapha Sadouki

Preface

This book, *Porous Fluids - Advances in Fluid Flow and Transport Phenomena in Porous Media*, discusses fluid flow and transport phenomena as encountered in civil and environmental engineering, petroleum and mathematical aspects, as well as chemical and geothermal engineering. It also covers the study of multiphase flow through porous media and introduces new methods. Portions of the book are devoted to interpretations of experimental results in this area and directions for future research. Written by authoritative experts in the field, this book can be used by applied mathematicians and engineers, especially those working in the area of porous media.

Vallampati Ramachandra Prasad,
Professor,
Department of Mathematics,
School of Advanced Sciences,
Vellore Institute of Technology,
Vellore, Tamil Nadu, India

Section 1

Multiscale Modelling in Porous Media

Chapter 1

Porous Flow with Diffuse Interfaces

Paul Papatzacos

Abstract

This chapter presents a model developed by the author, in publications dated from 2002 to 2016, on flow in porous media assuming diffuse interfaces. It contains five sections. Section 1 is an Introduction, tracing the origin of the diffuse interface formalism. Section 1 also presents the traditional compositional model, pointing out its emphasis on phases and questioning the concept of relative permeabilities. Section 2 presents the mass, momentum, and energy balance equations, for a multicomponent continuous fluid, in their most general form, at the pore level. The existence of constitutive equations with phase-inducing terms is mentioned, but the equations are not introduced at this level, and phases are not an explicit concern. Section 3 is about the averaging of the pore level equations inside a region containing many pores. There is no explicit mention of phases and therefore not of relative permeabilities. Section 4 is the technical basis from which the constitutive equations of the model arise, and it is shown that many models can exist. Section 5 introduces constitutive equations and presents a *minimal model* for multicomponent, multiphase, and thermal flow in neutrally wetting porous media, i.e., a model with a minimal amount of phenomenological parameters.

Keywords: flow in porous media, Marle averaging, diffuse interface, multiphase flow, phase segregation, relative permeabilities

1. Introduction

The model presented in this chapter was developed in Refs. [1–3]. Ref. [1] covers the one-component two-phase case, Ref. [2] is a generalisation to an arbitrary number of components and three phases (two liquids, one gas), Ref. [3] generalises to variable temperature. All three show applications to neutrally wetting media. The way wetting can be accounted for is discussed in general terms in Ref. [1]. A partial practical implementation, valid for incomplete wetting, is suggested in Ref. [4] but a fully satisfactory solution that accounts for total wetting (or capillary condensation in adsorption terms) is still a matter for further research.

The diffuse interface theory was initiated in 1893 in a paper by van der Waals (see the translation by Rowlinson [5]) where he proposes to replace the old assumption of a surface between phases by the assumption of a continuous transition inside a thin interphase region, where certain quantities, notably the density, vary continuously. The core of his theory consists of a Helmholtz function modified by the addition of a term proportional to the squared gradient of the density, thus accounting for the energy stored in the region. *In the model presented in the following pages, the van der Waals theory is introduced at the upscaled level.* See Section 4. Generalising the van der Waals expression to ν chemical components [2], one obtains

$$\mathcal{F}^{\mathrm{F}} = \mathcal{F}^{\mathrm{bF}} + \sum_{\alpha=1}^{\nu} \frac{1}{2} \Lambda^{\alpha} |\nabla R^{\alpha}|^{2}, \tag{1}$$

where R^{α} is the upscaled density of fluid component α, $\mathcal{F}$ is the Helmholtz function per unit volume, the superscript F referring to the fluid, while bF means "bulk fluid" and refers to fluid regions that are far away from interphase layers. The Λ^{α} are constants (Assumption A7, Appendix 1). For later reference $\mathcal{F}^{\mathrm{bF}}$ and its differential are

$$\mathcal{F}^{\mathrm{bF}} = -P^{\mathrm{bF}} + \sum_{\alpha=1}^{\nu} \mathcal{C}^{\mathrm{bF}\alpha} R^{\alpha}, \qquad \mathrm{d}\mathcal{F}^{\mathrm{bF}} = -\mathcal{S}^{\mathrm{bF}} \mathrm{d}T + \sum_{\alpha=1}^{\nu} \mathcal{C}^{\mathrm{bF}\alpha} \mathrm{d}R^{\alpha}, \tag{2}$$

where P^{bF} is the pressure in the bulk fluid, $\mathcal{S}^{\mathrm{bF}}$ the entropy per unit volume, and $\mathcal{C}^{\mathrm{bF}\alpha}$ the chemical potential of component α, divided by its molar mass (the C^{α}/M^{α} of Refs. [2, 3]).

The van der Waals paper was followed, in 1901, by a paper by Korteweg [6] about the equations of motion of a fluid with large but continuous density changes, where Korteweg showed that, for such a fluid, the usual scalar pressure must be replaced by a symmetric second order tensor.

The van der Waals-Korteweg papers were apparently forgotten, then rediscovered in the nineteen seventies, the diffuse-interface method being introduced as a novel way to solve fluid mechanics problems in two-phase flow. For a review, see [7]. While van der Waals and Korteweg assumed that the gradients in Eq. (1) are small, modern advances have shown that this limitation can be lifted [7, 8].

Ref. [8] is, to the author's knowledge, the first formulation of flow in porous media with diffuse interfaces.[1] The purpose with such a novel formulation is to avoid some of the known weaknesses of the traditional compositional models treating multiple phase problems. The mathematical core of the compositional models used in reservoir engineering, consists of equations expressing mass balance for each chemical component. The distribution of the phases in the reservoir is essential to the formulation, and is determined at the beginning of each time step. A component, present in a phase, is transported with the Darcy velocity of the phase. Each phase-dependent Darcy velocity is written with a permeability that is modified by a multiplicative factor, the relative permeability. The above description emphasises two of the weaknesses of the traditional models. The first is the assumption that well-defined phases exist at all times. The second is the use of relative permeabilities, a concept that is "seriously questioned", as expressed by Adler and Brenner in a 1988 paper [9]. Concerning the understanding of relative permeabilities in the framework of the model presented here, see the comments following Eq. (100) below.

The mathematical core of the model (actually a family of models) to be presented consists of mass balance equations, one per chemical component, of a momentum balance equation, and of an entropy balance equation. The

[1] The author first became aware of this paper in 2015. There is one important difference with what is presented in the pages that follow. See the footnote in Section 4.

thermodynamical description of the fluid mixture involved is part of the core. The central purpose is to calculate the component densities, and other characteristic quantities such as fluid velocity and fluid temperature, as functions of space and time. If the approximation of constant temperature is valid, only the mass and momentum balance equations are necessary.

Phases (and thereby relative permeabilities) do not take part in the formulation. They result from the solutions of the model equations, and are detected by rapid variations of densities, and by regions of approximately uniform densities. They can be shown to exist in static equilibrium or steady state dynamical situations [1–3].

A *minimal model* is presented in Section 5, consisting of a minimal amount of parameters.

A note on the appendices: Some concepts are grouped in appendices for easy reference. Appendix 1, for example, lists all the assumptions the model is built on.

A note on wetting: The problem of accounting for the wetting properties of the pore surface remains to be solved. There are two approaches to the problem: through boundary conditions to the Navier–Stokes equations at pore level, or through the theory of adsorption at the upscaled Darcy level. The first approach has been used in the publications considered here, and it is explained in Ref. [10] that the diffuse interface theory presented in Ref. [1] is consistent for neutral wetting, i.e., for pore level wetting angles around 90°: see Appendix 1, Assumption A0. Refs. [2, 3] assume this limitation.

A note on notation: Right-handed Cartesian coordinates (x_1, x_2, x_3) are assumed, the plane (x_1, x_2) being horizontal, and axis x_3 pointing upwards. Any vector $\mathbf{A}$ has components A_k where k is 1, 2, or 3. In addition, the notation $\partial_t = \partial/\partial t$ and $\partial_k = \partial/\partial x_k$ is used. The summation convention applies to latin indexes i, j, k, l: an index that is repeated in a term (as in $A_j B_j$, or C_{kk}) indicates summation ($\sum_{j=1}^{3} A_j B_j$, or $\sum_{k=1}^{3} C_{kk}$). Note that an index that is repeated, but not in the same term (as in $A_i = B_i$), means that the expression is valid for all values of the index in the set $\{1, 2, 3\}$. Symbols are otherwise defined when introduced.

2. Pore level equations

The fluid is a mixture of ν chemical components, it is continuous, and if phases exist, there are interphase regions where quantities vary continuously, possibly rapidly. With assumptions A1 to A3 (Appendix 1) the balance equations for mass, momentum, and energy, are written below in a most general manner:

$$\partial_t \rho^\alpha + \partial_k \left(\rho^\alpha v_k + i_k^\alpha \right) = 0, \qquad (\alpha = 1, \dots, \nu), \tag{3}$$

$$\partial_t (\rho v_i) + \partial_k \left(\rho v_i v_k - t_{ki} \right) = \rho f_i, \tag{4}$$

$$\partial_t \varepsilon + \partial_k \left(\varepsilon v_k + j_k - t_{ki} v_i \right) = \rho f_k v_k. \tag{5}$$

For the theoretical basis of these equations see chapter 11 of the book by Hirschfelder *et al* [11]. Greek superscripts indicate the species so that ρ^α is the mass density of component α, while ρ is the total density:

$$\rho = \sum_{\alpha=1}^{\nu} \rho^\alpha. \tag{6}$$

Further, $\mathbf{v}$ is the local velocity, defined as the total momentum divided by the total mass, both per volume; $\boldsymbol{\iota}^\alpha$ is the non-convective mass current of component α, with the property

$$\sum_{\alpha=1}^{\nu} \boldsymbol{\iota}^{\alpha} = \mathbf{0}, \tag{7}$$

$\boldsymbol{J}$ is the non-convective energy current, and t_{ij} is the stress tensor; $\varepsilon = u + \frac{1}{2}\rho v^2$ is the total energy, u is the internal energy, both per unit volume, and $\mathbf{f}$ is an external force per unit mass of fluid, thus being species-independent. In the case of gravity

$$\mathbf{f} = \mathbf{g}, \quad \mathbf{g} = (0, 0, -g), \quad g = 9.81\,\mathrm{m/s^2}. \tag{8}$$

It is convenient, for later use, to introduce the gravitational potential

$$W = -\mathbf{g}\cdot\mathbf{x}, \quad \text{whereby} \quad f_i = -\partial_i W. \tag{9}$$

The symmetry $t_{ij} = t_{ji}$ is assumed below. It is reminded that it expresses angular momentum balance for a fluid where no other torques exist than the one due to the external force $\mathbf{f}$, and the one due the surface stress $\mathbf{n}\cdot\mathbf{t}$ acting on every surface element where the normal vector is $\mathbf{n}$.

Within the van der Waals theory, one expects that ι_k^{α}, J_k, and t_{ki}, contain terms whose magnitudes are important in the interphase regions, but are otherwise negligible. No constitutive equations are introduced at the pore level, but it is mentioned for later reference, that for a simple one-phase fluid,

$$t_{ij} = -p\delta_{ij} + \theta_{ij}, \tag{10}$$

where p is the pressure and θ_{ij} is the viscous stress tensor (symmetric and linear in the gradients of the v_i, with coefficients of shear and bulk viscosity).

The upscaling, i.e., the averaging over many pores, is done in the next section by the method due to Marle [12]. This method assumes that the physical quantities that appear in the balance equations above are treated as distributions [12, 13], the underlying reason being that such quantities are discontinuous, and that one needs to average their partial derivatives. Taking ρ as an example, it is (i) a fluid density in a pore, (ii) a rock density in the rock matrix, (iii) undefined on the pore surface, and one needs to average $\partial_t\rho$ and $\partial_k\rho$.

The physical quantities appearing in Eqs. (3) to (5) are listed in the first line of **Table 1**. Their values in the pores, or in the rock, are denoted with a superscript F, or S, as shown in the second and third lines of the table. In the third line, a missing entry indicates non-existence, the first two 0-values indicate no material transport in the solid, the third zero value follows from Assumption A4 (Appendix 1): as shown by Marle [12], this assumption implies that the momentum balance equations in the solid and on the pore surface have the form $\mathbf{0} = \mathbf{0}$. It is important to keep in mind, especially when averaging, that any quantity with an F superscript is

Generic	ρ^{α}	ρ	v_k	ι_k^{α}	t_{ki}	f_k	u	j_k
In fluid	$\rho^{F\alpha}$	ρ^F	v_k^F	$\iota_k^{F\alpha}$	t_{ki}	f_k	u^F	j_k^F
In solid		ρ^S	0	0		0	u^S	j_k^S

Table 1.
First line, left of the vertical: quantities appearing in Eqs. (3) and (4); right of the vertical: quantities only appearing in Eq. (5). Second and third lines: notation when specialising to the fluid and the solid. Concerning the missing entries and the three 0-values, see beginning of paragraph containing Eq. (11).

equal to 0 on the pore surface (denoted by Σ) and in the solid; likewise, any quantity with superscript S is equal to 0 on Σ and in the fluid. The quantities in the first line of the table are not defined on Σ, but limit values are, as in

$$\left(\mathbf{v}^{\mathrm{F}}\right)^{\Sigma} = \mathbf{0}, \tag{11}$$

expressing that the fluid velocity vanishes on Σ. Expressions of the type $(X)^{\Sigma}$ for some X, used in this section and in the next, define the limit of the quantity X at Σ, along the line carrying the normal to Σ pointing towards the fluid. The existence of the normal implies some idealisation of the pore surface.

Note finally that, according to Eq. (6),

$$\rho^{\mathrm{F}} = \sum_{\alpha=1}^{\nu} \rho^{\mathrm{F}\alpha}. \tag{12}$$

As a preliminary to upscaling, Eqs. (3)–(5) are now explicitly written in terms of distributions, using the notation of Appendix 2 where A is replaced by Fα or F, as appropriate, while B is replaced by S. As an example, $\rho^{\alpha}(\mathbf{x}, t)$, is a distribution depending on space $\mathbf{x}$ and time t. It is continuous in time but not in space, being equal to $\rho^{\mathrm{F}\alpha}(\mathbf{x}, t)$ if $\mathbf{x}$ is inside a pore and to $\rho^{\mathrm{S}}(\mathbf{x}, t)$ if $\mathbf{x}$ is in the rock. It is not defined when $\mathbf{x}$ is on Σ but we assume that $\left(\rho^{\mathrm{F}\alpha}\right)^{\Sigma}$ and $\left(\rho^{\mathrm{S}}\right)^{\Sigma}$ exist.

The generalised mass balance equation is obtained from Eq. (3), using Eqs. (112), (113), and (11):

$$\{\partial_t \rho^{\alpha}\} + \left\{\partial_k\left(\rho^{\alpha} v_k + i_k^{\alpha}\right)\right\} + \left[\left(\imath^{\alpha} \cdot \mathbf{n}\right)^{\Sigma}\right]\delta_{\Sigma} = 0, \qquad (\alpha = 1, \dots, \nu). \tag{13}$$

From this equation one now gets, using **Table 1**, the equations that are separately valid inside the pores, on Σ, and in the solid:

$$3\partial_t \rho^{\mathrm{F}\alpha} + \partial_k\left(\rho^{\mathrm{F}\alpha} v_k^{\mathrm{F}} + i_k^{\mathrm{F}\alpha}\right) = 0, \quad (\alpha = 1, \dots, \nu), \quad (\text{in } F) \tag{14}$$

$$\left[\left(\imath^{\mathrm{F}\alpha} \cdot \mathbf{n}\right)^{\Sigma}\right]\delta_{\Sigma} = 0, \qquad (\alpha = 1, \dots, \nu), \quad (\text{on } \Sigma) \tag{15}$$

$$\partial_t \rho^{\mathrm{S}} = 0. \quad (\text{in } S) \tag{16}$$

Turning to the generalised momentum balance equation, one must account for Assumption A4 (Appendix 1), implying that the momentum balance equations in the solid and on the pore surface have the form $\mathbf{0} = \mathbf{0}$. Inside the pores, one simply re-writes Eq. (4), with the superscript F on the density and velocity:

$$\partial_t\left(\rho^{\mathrm{F}} v_i^{\mathrm{F}}\right) + \partial_k\left(\rho^{\mathrm{F}} v_k^{\mathrm{F}} v_i^{\mathrm{F}} - t_{ki}\right) = \rho^{\mathrm{F}} f_i, \quad (\text{in } F). \tag{17}$$

The generalised energy balance equation is obtained from Eq. (5), using Eqs. (112), (113), and (11):

$$\begin{aligned} &\left\{\partial_t\left(u + \frac{1}{2}\rho|\mathbf{v}|^2\right)\right\} + \left\{\partial_k\left(u v_k + \frac{1}{2}\rho|\mathbf{v}|^2 v_k + j_k - t_{ki} v_i\right)\right\} \\ &+ \left[\left(\jmath^{\mathrm{F}} \cdot \mathbf{n}\right)^{\Sigma} - \left(\jmath^{\mathrm{S}} \cdot \mathbf{n}\right)^{\Sigma}\right]\delta_{\Sigma} = f_k \rho v_k. \end{aligned} \tag{18}$$

Using **Table 1**, one then obtains:

$$\partial_t\left(u^{\mathrm{F}}+\frac{1}{2}\rho^{\mathrm{F}}\left|\mathbf{v}^{\mathrm{F}}\right|^2\right)+\partial_k\left(u^{\mathrm{F}}v_k^{\mathrm{F}}+\frac{1}{2}\rho^{\mathrm{F}}\left|\mathbf{v}^{\mathrm{F}}\right|^2v_k^{\mathrm{F}}+J_k^{\mathrm{F}}-t_{ki}v_i^{\mathrm{F}}\right)=f_k\rho^{\mathrm{F}}v_k^{\mathrm{F}},\quad(\text{in F}) \tag{19}$$

$$\left[\left(\boldsymbol{J}^{\mathrm{F}}\cdot\mathbf{n}\right)^{\Sigma}-\left(\boldsymbol{J}^{\mathrm{S}}\cdot\mathbf{n}\right)^{\Sigma}\right]\delta_{\Sigma}=0,\quad(\text{on }\Sigma) \tag{20}$$

$$\partial_t u^{\mathrm{S}}+\partial_k J_k^{\mathrm{S}}=0,\quad(\text{in S}). \tag{21}$$

3. Averaging

The Marle averaging process [12] is followed in all essentials, except in the assumption that well-defined phases exist, separated by interphase surfaces. The averaging volume is a sphere of radius r, large when compared to a pore radius, small when compared to a linear dimension of the reservoir. A C^{∞} function $m(\mathbf{x})$ is introduced, somewhat flat around $|\mathbf{x}|=0$ and equal to zero for $|\mathbf{x}|\geq r$, normalised so that its integral over all space is equal to 1. (See Eq. (17) in Ref. [12] for an example of such a function.) Given any function of space and time, $f(\mathbf{x},t)$, its average $F(\mathbf{x},t)$ is obtained by the convolution

$$F(\mathbf{x},t)\equiv(f*m)(\mathbf{x},t)=\int_{\mathbb{R}^3}f(\mathbf{y},t)m(\mathbf{x}-\mathbf{y})\mathrm{d}\mathbf{y}, \tag{22}$$

where the integration is over all of space. The convolution ensures that F is C^{∞} [12].

The averaged balance equations are differential equations in the averaged quantities (averaged densities, velocities, ...). These equations are established by a three-step process. Step 1: the generalised equations for mass, momentum, and energy balance are each in turn convoluted with m. Step 2: the following rules [12] are applied, allowing to take the differential operators out of the averaging convolutions:

$$\left\{\partial_t f^{\mathrm{Z}}\right\}*m=\partial_t\left(f^{\mathrm{Z}}*m\right), \tag{23}$$

$$\left\{\partial_i f^{\mathrm{Z}}\right\}*m=\partial_i\left(f^{\mathrm{Z}}*m\right)-\varepsilon^{\mathrm{Z}}\left[\left(f^{\mathrm{Z}}\right)^{\Sigma}n_i\delta_{\Sigma}\right]*m, \tag{24}$$

where $f^{\mathrm{Z}}\equiv f^{\mathrm{Z}}(\mathbf{x},t)$ and Z is F, Fα, or S; ε^{Z} is 1 if Z is F or Fα, -1 if Z is S. Eq. (11) is also applied at step 2. Step 3: the remaining convolutions are used to define averaged quantities, where the following constraints should be obeyed: (i) except for the definition of porosity, the way to define the averages is suggested by the equations obtained after completion of step 2; (ii) the averaged equations have essentially the same forms as Eqs. (3) to (5).

Differential equations in the averaged quantities result from the three steps. It is shown below that the mass and momentum balance equations should be treated together, and that the energy equation can be treated as an addition.

The averaging of the mass and momentum balance equations follows. Steps 1 and 2 are applied to Eqs. (14) to (17), and lead to:

$$\partial_t\left(\rho^{\mathrm{F}\alpha}*m\right)+\partial_k\left(\left(\rho^{\mathrm{F}\alpha}v_k^{\mathrm{F}}+\iota_k^{\mathrm{F}\alpha}\right)*m\right)=\left(\left(\boldsymbol{\iota}^{\mathrm{F}\alpha}\cdot\mathbf{n}\right)^{\Sigma}\delta_{\Sigma}\right)*m, \tag{25}$$

$$\left(\left(\boldsymbol{\iota}^{\mathrm{F}\alpha}\cdot\mathbf{n}\right)^{\Sigma}\delta_{\Sigma}\right)*m=0, \tag{26}$$

$$\partial_t \left(\rho^{\mathrm{S}} * m \right) = 0, \tag{27}$$

$$\partial_t \left(\left(\rho^{\mathrm{F}} v_i^{\mathrm{F}} \right) * m \right) + \partial_k \left(\left(\rho^{\mathrm{F}} v_k^{\mathrm{F}} v_i^{\mathrm{F}} - t_{ki} \right) * m \right) = -\left(\left(n_k t_{ki} \right)^{\Sigma} \delta_{\Sigma} \right) * m + f_i \left(\rho^{\mathrm{F}} * m \right). \tag{28}$$

Note that the last term on the right-hand side of the last equation originally is $(\rho^{\mathrm{F}} f_i) * m$, but f_i can be taken out of the convolution integral because of Assumption A5 (Appendix 2). The definitions of averaged (or upscaled) quantities now follow (step 3). Porosity Φ is defined first, then follow definitions suggested by the convolution operations in the equations above.

Porosity Let a function $\chi(\mathbf{x})$ be 1 when $\mathbf{x}$ is in F, and let it be 0 otherwise. Then

$$\chi * m = \Phi. \tag{29}$$

Note that, according to definition (22), Φ can depend on $\mathbf{x}$.

Species adsorption The left-hand side of Eq. (26) defines the amount, $K^{\Sigma\alpha}$, of component α adsorbed at the pore surface:

$$\left(\left(\boldsymbol{\iota}^{\mathrm{F}\alpha} \cdot \mathbf{n} \right)^{\Sigma} \delta_{\Sigma} \right) * m = K^{\Sigma\alpha}. \tag{30}$$

Density of solid Eq. (27) suggests defining the solid density R^{S} by:

$$\rho^{\mathrm{S}} * m = (1 - \Phi) R^{\mathrm{S}}. \tag{31}$$

Density of component α The first term on the left-hand side of Eq. (25) suggests defining the density R^α of fluid component α by

$$\rho^{\mathrm{F}\alpha} * m = R^\alpha. \tag{32}$$

Note that Eq. (12) implies that the averaged total fluid density, R, is

$$R = \sum_{\alpha=1}^{\nu} R^\alpha = \rho^{\mathrm{F}} * m. \tag{33}$$

Fluid velocity The first term on the lef-hand side of Eq. (28) has the convolution of a product of two term, where the average of one of them is known from Eqs. (32) and (33). The averaged fluid velocity, denoted V_i (without the F superscript since there is no velocity in S or on Σ) is then defined by

$$\left(\rho^{\mathrm{F}} v_i^{\mathrm{F}} \right) * m = \Phi R V_i. \tag{34}$$

Diffusive mass current of component α The second convolution on the left-hand side of Eq. (25) is used to define the upscaled diffusive current in the fluid, denoted I_k^α (without the F superscript since there are no diffusive currents in S or on Σ). It is not defined as the average of $\iota_k^{\mathrm{F}\alpha}$ because both pore level effects of convection and diffusion contribute to it [12]. Keeping in mind the constraint that the averaged equations should have the same form as the original ones, I_k^α is defined by

$$\left(\rho^{\mathrm{F}\alpha} v_k^{\mathrm{F}} + \iota_k^{\mathrm{F}\alpha} \right) * m = \Phi R^\alpha V_k + \Phi I_k^\alpha. \tag{35}$$

Note that, summing this equation over α from 1 to ν, and using previous results, one gets

$$\sum_{\alpha=1}^{\nu} I_k^{\alpha} = 0. \tag{36}$$

Stress tensor The second convolution on the left-hand side of Eq. (28) suggests defining T_{ki}, the upscaled version of t_{ki}, by

$$\left(\rho^{\mathrm{F}} v_k^{\mathrm{F}} v_i^{\mathrm{F}} - t_{ki}\right) * m = \Phi R V_i V_k - T_{ki}. \tag{37}$$

Note that $t_{ki} = t_{ik}$ implies $T_{ki} = T_{ik}$.

Frictional force per unit volume The first convolution on the right-hand side of Eq. (28) suggests defining the frictional force per unit volume $\mathfrak{F}_i^{\mathrm{F}}$ by

$$\left((n_k t_{ki})^{\Sigma} \delta_{\Sigma}\right) * m = \mathfrak{F}_i^{\mathrm{F}}. \tag{38}$$

The upscaled mass and momentum balance equations now follow, in the following order: mass balance for the solid, mass balance at the pore surface, mass balance for the fluid in the pores, and momentum balance for the fluid in the pores:

$$\partial_t\left[(1-\Phi)R^{\mathrm{S}}\right] = 0, \tag{39}$$

$$K^{\Sigma\alpha} = 0, \quad (\alpha = 1 \dots \nu), \tag{40}$$

$$\partial_t(\Phi R^{\alpha}) + \partial_k\left(\Phi R^{\alpha} V_k + \Phi I_k^{\alpha}\right) = 0, \quad (\alpha = 1 \dots \nu), \tag{41}$$

$$\partial_t(\Phi R V_i) + \partial_k(\Phi R V_k V_i - T_{ki}) = \mathfrak{F}_i^{\mathrm{F}} + \Phi R f_i. \tag{42}$$

The first equation states that porosity and solid density do not vary with time (consistently with Assumption 4 (Appendix 2) but can vary in space. The second equation states that adsorption is negligibly small, for any component, consistent with Assumption A0 (Appendix 2). The remaining two equations determine the ν component densities and the three velocity components. This means that the components of the diffusive mass current, of the stress tensor, and of the frictional force must be provided. (Constitutive equations).

The averaging of the energy balance equations now follows. Steps 1 and 2 are applied to Eqs. (19) to (21), giving:

$$\begin{aligned}&\partial_t\left[\left(u^{\mathrm{F}} + \frac{1}{2}\rho^{\mathrm{F}}|\mathbf{v}^{\mathrm{F}}|^2\right) * m\right] + \partial_k\left[\left(u^{\mathrm{F}} v_k^{\mathrm{F}} + \frac{1}{2}\rho^{\mathrm{F}}|\mathbf{v}^{\mathrm{F}}|^2 v_k^{\mathrm{F}} + J_k^{\mathrm{F}} - t_{ki} v_i^{\mathrm{F}}\right) * m\right]\\ &= \left[\left(J^{\mathrm{F}}\cdot\mathbf{n}\right)^{\Sigma}\delta_{\Sigma}\right] * m + f_k\left(\rho^{\mathrm{F}} v_k^{\mathrm{F}} * m\right),\end{aligned} \tag{43}$$

$$\left(\left(J^{\mathrm{F}}\cdot\mathbf{n}\right)^{\Sigma}\delta_{\Sigma}\right) * m - \left(\left(J^{\mathrm{S}}\cdot\mathbf{n}\right)^{\Sigma}\delta_{\Sigma}\right) * m = 0, \tag{44}$$

$$\partial_t\left(u^{\mathrm{S}} * m\right) + \partial_k\left(J_k^{\mathrm{S}} * m\right) = -\left[\left(J^{\mathrm{S}}\cdot\mathbf{n}\right)^{\Sigma}\delta_{\Sigma}\right] * m. \tag{45}$$

As in the case of Eq. (28), Assumption A4 (Appendix 2) has been used to take f_k out of the convolution on the right-hand side of Eq: (43). Six definitions are introduced below, built on the definitions that were introduced in connection with the averaging of the mass and momentum balance equations. An underscore indicates the defined quantity.

Internal energy per unit volume of solid

$$u^{\mathrm{S}} * m = (1 - \Phi)\underline{\mathcal{U}^{\mathrm{S}}}, \tag{46}$$

Internal energy current of solid

$$j_k^{\mathrm{S}} * m = \underline{J_k^{\mathrm{S}}}, \tag{47}$$

Solid to fluid energy transfer

$$\left(\left(\boldsymbol{j}^{\mathrm{S}} \cdot \mathbf{n} \right)^{\Sigma} \delta_{\Sigma} \right) * m = \underline{Q_{\mathrm{S} \to \mathrm{F}}}, \tag{48}$$

Fluid to solid energy transfer

$$\left(\left(\boldsymbol{j}^{\mathrm{F}} \cdot \mathbf{n} \right)^{\Sigma} \delta_{\Sigma} \right) * m = -\underline{Q_{\mathrm{F} \to \mathrm{S}}}, \tag{49}$$

Internal energy per unit volume of fluid

$$\left(u^{\mathrm{F}} + \frac{1}{2} \rho^{\mathrm{F}} \left| \mathbf{v}^{\mathrm{F}} \right|^2 \right) * m = \Phi \underline{\mathcal{U}^{\mathrm{F}}} + \frac{1}{2} \Phi R |\mathbf{V}|^2, \tag{50}$$

Internal energy current of fluid

$$\left(u^{\mathrm{F}} v_k^{\mathrm{F}} + \frac{1}{2} \rho^{\mathrm{F}} \left| \mathbf{v}^{\mathrm{F}} \right|^2 v_k^{\mathrm{F}} + j_k^{\mathrm{F}} - t_{ki} v_i^{\mathrm{F}} \right) * m = \Phi \mathcal{U}^{\mathrm{F}} V_k + \frac{1}{2} \Phi R |\mathbf{V}|^2 V_k + \underline{J_k^{\mathrm{F}}} - T_{ki} V_i. \tag{51}$$

Using these definitions in Eqs. (43) to (45) one obtains:

$$\begin{aligned} &\partial_t \left(\Phi \mathcal{U}^{\mathrm{F}} + \frac{1}{2} \Phi R |\mathbf{V}|^2 \right) + \partial_k \left(\Phi \mathcal{U}^{\mathrm{F}} V_k + \frac{1}{2} \Phi R |\mathbf{V}|^2 V_k + J_k^{\mathrm{F}} - T_{ki} V_i \right) \\ &= -Q_{\mathrm{F} \to \mathrm{S}} + \Phi R V_k f_k. \quad (\text{in F}) \end{aligned} \tag{52}$$

$$Q_{\mathrm{F} \to \mathrm{S}} + Q_{\mathrm{S} \to \mathrm{F}} = 0. \quad (\text{on } \Sigma) \tag{53}$$

$$\partial_t \left[(1 - \Phi) \mathcal{U}^{\mathrm{S}} \right] + \partial_k J_k^{\mathrm{S}} = -Q_{\mathrm{S} \to \mathrm{F}}. \quad (\text{in S}). \tag{54}$$

Eq. (52) contains a redundancy in the form of a balance equation for kinetic energy. This equation can be obtained directly by multiplying both sides of Eq. (42) with V_i, summing over i, and using Eq. (41). Subtracting the equation thus obtained from Eq. (52), one gets

$$\partial_t \left(\Phi \mathcal{U}^{\mathrm{F}} \right) + \partial_k \left(\Phi \mathcal{U}^{\mathrm{F}} V_k + J_k^{\mathrm{F}} \right) = -Q_{\mathrm{F} \to \mathrm{S}} - \mathfrak{F}_k^{\mathrm{F}} V_k + T_{ji} \partial_j V_i. \tag{55}$$

4. Basis for constitutive equations

In practical application, the upscaled balance equations will be used to calculate, primarily, the densities R^{α}, the velocity components V_i and the temperature T. Assuming that the approximation of constant T is valid, one needs only focus on the mass and momentum balance equations, (39) to (42). One sees in this case that one needs expressions for the I_k^{α}, the T_{ij}, and the $\mathfrak{F}_i$ in terms of T and the R^{α}. If the approximation of constant temperature is not valid, one must first distinguish between the temperatures of the solid and the fluid, and one needs an equation for the energy transfer in case of a temperature difference. One introduces the simplifying Assumption A6 (see Ref. [12] and Appendix 1), from which it follows that just one additional equation is needed for calculating $T(\mathbf{x},t)$. Such an equation usually describes the evolution of either total energy or total entropy. In either case, expressions are needed for the currents J_k^{S} and J_k^{F}. (Expressions for the energy transfers $Q_{\mathrm{S}\to\mathrm{F}}$ and $Q_{\mathrm{F}\to\mathrm{S}}$ are unnecessary since they cancel when taking the sum of the solid and fluid energies). Most of what is needed is obtained in Section 5 by applying the theory of irreversible processes, starting from the evolution equation for entropy, although it seems that some preliminary work is unavoidable to directly obtain an expression for the pressure tensor P_{ij} that replaces the usual scalar pressure.

The derivation of the pressure tensor is given below, followed by the derivation of the evolution equation for the total entropy. *It is essential to use expression (1) in both derivations.*[2]

4.1 The pressure tensor

One considers the upscaled fluid, consisting of a mixture of ν components in a container with surface $\partial\Omega$ and volume Ω, inside a large bath at uniform and constant temperature T. One looks for conditions of equilibrium in the presence of gravity. The fluid has the Helmholtz free energy density given by Eq. (1), and it is assumed that the bounding surface is neutrally wetting so that there is no energy stored on $\partial\Omega$. The total energy stored in the fluid is

$$F = \int_{\Omega}\left[\mathcal{F}^{\mathrm{F}} + W\sum_{\alpha=1}^{\nu} R^{\alpha}\right] \mathrm{d}\Omega, \tag{56}$$

where W is the gravitational potential (see Eqs. (8) and (9)).

One now looks for the conditions the R^{α} satisfy when F is at its minimum, given that the total mass, $\int_{\Omega} R\,\mathrm{d}\Omega$, is constant. This is minimisation with constraint, a standard problem in variational calculus. It is easily found that

$$\frac{\partial\mathcal{F}^{\mathrm{bF}}}{\partial R^{\alpha}} + W - \kappa - \Lambda^{\alpha}\nabla^{2}R^{\alpha} = 0, \quad (\text{in} \quad \Omega\ ), \tag{57}$$

$$\mathbf{n}\cdot\nabla R^{\alpha} = 0, \quad (\text{on } \partial\Omega), \tag{58}$$

where κ is a Lagrange multiplier, possibly a function of T but not of the R^{α}. See Ref. [14] for the technique and the theorems involved: essentially, the expression on the left-hand side of Eq. (57) must be continuous. Note that Eq. (58) expresses the non-wetting property of the outer boundary: the density neither increases nor decreases along the normal.

The next step consists in multiplying Eq. (57) with $\partial_i R^{\alpha}$ and summing over α. Each term of the resulting equation can then be re-written as a gradient (first term), or as a sum of a gradient and the component of a force (sum of second and third terms), or as a sum of a gradient and a divergence (fourth term). The result is the following expression:

[2] This is where the present paper and Ref. [8] differ most.

$$\partial_k P_{ik} - f_i = 0. \tag{59}$$

where

$$P_{ik} = \left[P^{\mathrm{bF}} - \sum_{\alpha=1}^{\nu} \Lambda^\alpha R^\alpha \nabla^2 R^\alpha - \sum_{\alpha=1}^{\nu} \frac{1}{2} \Lambda^\alpha |\nabla R^\alpha|^2 \right] \delta_{ik} + \sum_{\alpha=1}^{\nu} \Lambda^\alpha (\partial_i R^\alpha)(\partial_k R^\alpha). \tag{60}$$

The first equation obviously generalises the classical $\nabla P = f$, and it is natural to generalise Eq. (10) by setting

$$T_{ij} = -\Phi P_{ij} + \Theta_{ij}. \tag{61}$$

T_{ij} and P_{ij} are symmetric, so that $\Theta_{ij} = \Theta_{ji}$. Eqs. (60) and (61) are used in the long calculation that lead to the evolution equation for the entropy. They are also used to get expressions (99) and (100) for the modified Darcy equation, where the presence of the Λ's is exclusively due to their presence in Eq. (60).

4.2 The entropy evolution equation

Eqs. (1) and (2), together with Eqs. (41), (54), (55), (60), and (61), are needed to get the entropy equation for the fluid.

Taking the differential of Eq. (1) and using Eq. (2), one obtains, keeping in mind the assumption that the Λ^α are constants,

$$\mathrm{d}\mathcal{F}^{\mathrm{F}} = -\mathcal{S}^{\mathrm{bF}} \mathrm{d}T + \sum_{\alpha=1}^{\nu} \mathcal{C}^{\mathrm{bF}\alpha} \mathrm{d}R^\alpha + \sum_{\alpha=1}^{\nu} \Lambda^\alpha (\partial_i R^\alpha) \mathrm{d}(\partial_i R^\alpha). \tag{62}$$

This equation leads immediately to two conclusions. There is no additional entropy, and no additional chemical potentials due to large density gradients, since: (i) $\mathcal{S}^{\mathrm{F}}$, defined as $\partial \mathcal{F}^{\mathrm{F}} / \partial T$, is equal to $\mathcal{S}^{\mathrm{bF}}$; (ii) $\mathcal{C}^{\mathrm{F}\alpha}$, defined as $\partial \mathcal{F}^{\mathrm{F}} / \partial R^\alpha$, is equal to $\mathcal{C}^{\mathrm{bF}\alpha}$. The differential of $\mathcal{F}^{\mathrm{F}}$ can now be re-written with a simpler notation:

$$\mathrm{d}\mathcal{F}^{\mathrm{F}} = -\mathcal{S}^{\mathrm{F}} \mathrm{d}T + \sum_{\alpha=1}^{\nu} \mathcal{C}^{\mathrm{F}\alpha} \mathrm{d}R^\alpha + \sum_{\alpha=1}^{\nu} \Lambda^\alpha (\partial_i R^\alpha) \mathrm{d}(\partial_i R^\alpha). \tag{63}$$

To obtain $\mathrm{d}\mathcal{S}^{\mathrm{F}}$ in terms of $\mathrm{d}\mathcal{U}^{\mathrm{F}}$ and $\mathrm{d}\mathcal{F}^{\mathrm{F}}$ one differentiates $\mathcal{U}^{\mathrm{F}} = T\mathcal{S}^{\mathrm{F}} + \mathcal{F}^{\mathrm{F}}$, obtaining

$$T\mathrm{d}\mathcal{S}^{\mathrm{F}} = \mathrm{d}\mathcal{U}^{\mathrm{F}} - \sum_{\alpha=1}^{\nu} \mathcal{C}^{\mathrm{F}\alpha} \mathrm{d}R^\alpha - \sum_{\alpha=1}^{\nu} \Lambda^\alpha (\partial_k R^\alpha) \mathrm{d}(\partial_k R^\alpha). \tag{64}$$

This expression is now used to construct an evolution equation for the total entropy (including fluid and solid). Such an equation must be of the form

$$\partial_t \left[\Phi \mathcal{S}^{\mathrm{F}} + (1 - \Phi) \mathcal{S}^{\mathrm{S}} \right] + \partial_k \left(\Phi \mathcal{S}^{\mathrm{F}} V_k + P_k \right) = Q, \tag{65}$$

where $\mathcal{S}^{\mathrm{S}}$ is the entropy of the rock, P_k is a diffusive current, and Q is a source term.

Eq. (64) directly gives the two equations that follow, by replacing d with ∂_t (first equation), then with $V_i \partial_i$ (second equation):

$$T\partial_t \mathcal{S}^{\mathrm{F}} = \partial_t \mathcal{U}^{\mathrm{F}} - \sum_{\alpha=1}^{\nu} \mathcal{C}^{\mathrm{F}\alpha} \partial_t R^\alpha - \sum_{\alpha=1}^{\nu} \Lambda^\alpha (\partial_k R^\alpha) \partial_t (\partial_k R^\alpha). \tag{66}$$

$$T\partial_i \left(\mathcal{S}^{\mathrm{F}} V_i\right) = T\mathcal{S}^{\mathrm{F}} \partial_i V_i + V_i \partial_i \mathcal{U}^{\mathrm{F}} - \sum_{\alpha=1}^{\nu} \mathcal{C}^{\mathrm{F}\alpha} V_i \partial_i R^\alpha - V_i \sum_{\alpha=1}^{\nu} \Lambda^\alpha (\partial_k R^\alpha)(\partial_i \partial_k R^\alpha). \tag{67}$$

Since it is $\Phi\mathcal{S}^{\mathrm{F}}$ that is needed to get Eq. (65), one needs to multiply the two equations above with Φ and then commute Φ with ∂_t and ∂_i. The same commutations are necessary on the right-hand sides of the above equations, so as to get $\Phi\mathcal{U}^{\mathrm{F}}$ and thus allow the use of Eq. (55). Assumption A8 (Appendix 2) has been added to Assumption A4 so as to avoid the proliferation of $\partial_i\Phi$-terms in the minimal model to be presented. With this simplification in place, and remembering to replace T_{ij} in Eq. (55) with the expression obtained from Eqs. (60) and (61), one obtains, after elementary but somewhat long calculations:

$$\begin{aligned} T\left[\partial_t\left(\Phi\mathcal{S}^{\mathrm{F}}\right) + \partial_i\left(\Phi\mathcal{S}^{\mathrm{F}} V_i\right)\right] = &- \partial_k J_k^{\mathrm{F}} + Q_{\mathrm{S}\to\mathrm{F}} - \mathfrak{F}_k^{\mathrm{F}} V_k + \Theta_{ji}\partial_j V_i \\ &+ \Phi\sum_{\alpha=1}^{\nu} \mathcal{C}^{\mathrm{F}\alpha}\partial_i I_i^\alpha + \Phi\sum_{\alpha=1}^{\nu} \Lambda^\alpha (\partial_k R^\alpha)\partial_k\partial_i I_i^\alpha \\ &+ \Phi\sum_{\alpha=1}^{\nu} \Lambda^\alpha [R^\alpha(\partial_i V_i)\partial_k\partial_k R^\alpha + (\partial_k R^\alpha)\partial_k(R^\alpha \partial_i V_i)]. \end{aligned} \tag{68}$$

Keeping in mind that one is looking for an entropy equation of the form of Eq. (65), and in anticipation of using the methods of irreversible processes, one now considers each term, or group of terms, on the right-hand side above and writes it either as a divergence, or as the sum of a divergence and a scalar product. Most terms on the first line already have the required form. The general term in the first sum on the second line is easily transformed as required, the general term in the second sum also, although with somewhat more work. As to the third line, it is easily seen to be a sum of divergences. One then gets:

$$\begin{aligned} T\left[\partial_t\left(\Phi\mathcal{S}^{\mathrm{F}}\right) + \partial_i\left(\Phi\mathcal{S}^{\mathrm{F}} V_i\right)\right] = \; &\partial_k\left[-J_k^{\mathrm{F}} + \Phi\sum_{\alpha=1}^{\nu} A^\alpha I_k^\alpha + \Phi\sum_{\alpha=1}^{\nu} \Lambda^\alpha (\partial_k R^\alpha)\left(R^\alpha \partial_i V_i + \partial_i I_i^\alpha\right)\right] \\ &+ Q_{\mathrm{S}\to\mathrm{F}} - \mathfrak{F}_k^{\mathrm{F}} V_k + \Theta_{ji}\partial_j V_i - \Phi\sum_{\alpha=1}^{\nu} I_i^\alpha \partial_i A^\alpha, \end{aligned} \tag{69}$$

where $A^\alpha = \mathcal{C}^{\mathrm{F}\alpha} - \Lambda^\alpha \nabla^2 R^\alpha$. Note that A^α occurs twice on the right-hand side of Eq. (69), once as A^α, and once as $\partial_i A^\alpha$, both times multiplying I_k^α in a sum over α. Eq. (36) then implies that one can modify the above expression of A^α by the addition of any expression that does not depend on α. To conform with the notation of Ref. [2], and especially Ref. [3], one then sets

$$A^\alpha = \mathcal{C}^{\mathrm{F}\alpha} + W - \Lambda^\alpha \nabla^2 R^\alpha. \tag{70}$$

as the expression to substitute on the right-hand side of Eq. (69).

The entropy equation for the solid is much easier to obtain because of Assumption A4 (Appendix 2). Indeed, Eq. (64) is replaced by

$$T\mathrm{d}\mathcal{S}^{\mathrm{S}} = \mathrm{d}\mathcal{U}^{\mathrm{S}}, \tag{71}$$

which implies

$$T\partial_t\left[(1-\Phi)\mathcal{S}^{\mathrm{S}}\right] = \partial_t\left[(1-\Phi)\mathcal{U}^{\mathrm{S}}\right], \tag{72}$$

and, using (54):

$$T\partial_t\left[(1-\Phi)\mathcal{S}^{\mathrm{S}}\right] = -\partial_k J_k^{\mathrm{S}} - Q_{\mathrm{S}\to\mathrm{F}}. \tag{73}$$

Taking the sum of Eqs. (69) and (73) one sees that $Q_{\mathrm{S}\to\mathrm{F}}$ cancels, and that the right-hand side consists of a sum of scalar products and of a divergence, say $\partial_k K_k$. A last step remains because the left-hand side has a multiplicative factor T (see Eqs. (69) and (73)). Dividing both sides by T does not affect the scalar products but it modifies the divergence, at least if one assumes that T is variable, to $\partial_k K_k/T$. This is easily remedied by writing

$$\frac{1}{T}\partial_k K_k = \partial_k\left(\frac{K_k}{T}\right) + \frac{K_k}{T^2}\partial_k T. \tag{74}$$

One finally obtains

$$\partial_t\left[\Phi\mathcal{S}^{\mathrm{F}} + (1-\Phi)\mathcal{S}^{\mathrm{S}}\right] + \partial_k\left(\Phi\mathcal{S}^{\mathrm{F}}V_k + \frac{K_k}{T}\right) = Q, \tag{75}$$

where

$$K_k = J_k^{\mathrm{F}} + J_k^{\mathrm{S}} - \Phi\sum\nolimits_{\alpha=1}^{\nu}\Lambda^\alpha(\partial_k R^\alpha)\left(R^\alpha\partial_i V_i + \partial_i I_i^\alpha\right) - \Phi\sum_{\alpha=1}^{\nu} A^\alpha I_k^\alpha. \tag{76}$$

$$TQ = -(K_k/T)\partial_k T + \Theta_{ji}\partial_j V_i - \mathfrak{F}_k^{\mathrm{F}} V_k - \Phi\sum_{\alpha=1}^{\nu} I_k^\alpha\partial_k A^\alpha. \tag{77}$$

$Q_{\mathrm{F}\to\mathrm{S}}$ and $Q_{\mathrm{S}\to\mathrm{F}}$ do not appear in the entropy equation, and are thus not determinable inside the model. Similarly, the currents J_k^{F} and J_k^{S} only appear as summed, so that they are not determined individually inside the model. (See the next section.)

5. Constitutive equations and the minimal model

The source term in the entropy equation plays a central role in what follows. It has been written as a sum of scalar products. Each term of this sum is the scalar product of a force (explicit or generalised) and a current. $\mathfrak{F}_k^{\mathrm{F}}$ is an explicit force, while the gradient of some quantity (temperature, velocity component, ...) is a generalised force. The theory of irreversible processes states that linear relations exist (at least for processes not far from equilibrium), between forces and currents. The coefficients, called phenomenological coefficients, are parameters whose signs must be such that the source term cannot be negative, to ensure against a decrease of the entropy when the system is isolated. The linear relations just mentioned are constitutive equations, and it is implied that the phenomenological coefficients must be provided as input. For various reasons, some coefficients can be put equal to zero, one often recurring reason being the belief that they are negligible in the physical situation considered. Thus there is a family of models, each member being characterised by its set of non-zero phenomenological coefficients.

The previously mentioned minimal model (see the text after Eq. (67)), is the model that contains the least possible number of non-zero phenomenological coefficients. "Least possible" means that one must obey the constraints that exist for these coefficients (Onsager symmetry, isotropy, sign) and arbitrarily setting some of them equal to zero is not always possible. A certain amount of trial and error is also required to avoid unduly reducing the model's predictive power.

The usual vector notation is used for tensors of order 1, i.e., a bold faced letter is used when the subscript can be suppressed. Vector forces are here denoted **X** (with components X_i), and vector currents are denoted **Y** (with components Y_i), with a superscript to discriminate between the different currents and forces in an alphabetic order aa shown below. (Y has been chosen instead of the traditional J so as to avoid confusion with the currents related to the energy equations.) The usual tensor conventions are assumed for latin subscripts (see "A note on notation" in the Introduction). Concerning the second order tensors Θ_{ji} and $\partial_j V_i$, on the right-hand side of Eq. (77), one sets $\Delta_{ji} \equiv \partial_j V_i$ and, referring to Appendix 3, especially to Eq. (115), one writes

$$\Delta_{ij} = X^A \delta_{ij} + X^B_{ij} + \varepsilon_{ijk} \tilde{X}_k, \tag{78}$$

$$\Theta_{ij} = Y^A \delta_{ij} + Y^B_{ij}, \tag{79}$$

where X^A and X^B_{ij} are found by replacing Z by Δ in Eqs. (115). Note that Y^B_{ij}, being symmetric, has no antisymmetric part.

Referring now to the vectors on the right-hand side of Eq. (77), one introduces the following notation:

$$\begin{aligned} X^C_k &= -\partial_k T & Y^C_k &= K_k/T \\ X^D_k &= -\mathfrak{F}^{\mathrm{F}}_k & Y^D_k &= V_k \\ X^\alpha_k &= -\Phi \partial_k A^\alpha, & Y^\alpha_k &= I^\alpha_k. \end{aligned} \tag{80}$$

Using Eq. (116), one easily finds that

$$TQ = 3X^A Y^A + X^B_{ij} Y^B_{ij} + \mathbf{X}^C \cdot \mathbf{Y}^C + \mathbf{X}^D \cdot \mathbf{Y}^D + \sum_{\alpha=1}^{\nu} \mathbf{X}^\alpha \cdot \mathbf{Y}^\alpha. \tag{81}$$

Note that Q is a proper scalar since the source of entropy is independent of the coordinate system, and does not change sign when the coordinate system changes handedness. The right-hand side of Eq. (81) is then a sum of scalar products of proper tensors, the only pseudo vector, $\tilde{X}_k$, having dropped out.

Referring to the first sentence of this section, one writes the currents as linear combinations of the forces:

$$\begin{aligned} Y^A &= L^{AA} X^A + L^{AB}_{kl} X^B_{kl} + L^{AC}_k X^C_k + L^{AD}_k X^D_k + \sum_{\beta=1}^{\nu} L^{A\beta}_k X^\beta_k \\ Y^B_{ij} &= L^{BA}_{ij} X^A + L^{BB}_{ijkl} X^B_{kl} + L^{BC}_{ijk} X^C_k + L^{BD}_{ijk} X^D_k + \sum_{\beta=1}^{\nu} L^{B\beta}_{ijk} X^\beta_k \end{aligned} \tag{82}$$

$$Y_i^C = L_i^{CA} X^A + L_{ikl}^{CB} X_{kl}^B + L_{ik}^{CC} X_k^C + L_{ik}^{CD} X_k^D + \sum_{\beta=1}^{\nu} L_{ik}^{C\beta} X_k^\beta$$

$$Y_i^D = L_i^{DA} X^A + L_{ikl}^{DB} X_{kl}^B + L_{ik}^{DC} X_k^C + L_{ik}^{DD} X_k^D + \sum_{\beta=1}^{\nu} L_{ik}^{D\beta} X_k^\beta$$

$$Y_i^\alpha = L_i^{\alpha A} X^A + L_{ikl}^{\alpha B} X_{kl}^B + L_{ik}^{\alpha C} X_k^C + L_{ik}^{\alpha D} X_k^D + \sum_{\beta=1}^{\nu} L_{ik}^{\alpha\beta} X_k^\beta, \quad (\alpha = 1 \dots \nu).$$

In these expressions, the L are the phenomenological coefficients. They are independent of the generalised forces, but they can depend on the temperature and the component densities. They are tensors, their orders being equal to the number of subscripts. They obey the Onsager relations [11, 15]: any coefficient with n subscripts, denoted say, $L_{(n)}^{MN}$, obeys

$$L_{(n)}^{MN} = L_{(n)}^{NM}, \quad (M \neq N = A, B, C, D, 1, \dots, \nu), \tag{83}$$

where the n subscripts are the same but not necessarily in the same order. Details concerning subscripts are not needed in what follows.

5.1 The viscosity tensor Θ_{ij}

With hindsight, one knows that the upscaled viscosity tensor is not required, so that one is justified in setting equal to zero all the L's in the first two lines of the system of Eqs. (82), and also setting to zero all the L's related to the zeroed ones by the Onsager symmetry:

$$L^{AZ} = L^{BZ} = L^{ZA} = L^{ZB} = 0, \quad \text{for any Z}. \tag{84}$$

Then $Y^A = Y_{ij}^B = 0$, so that

$$\Theta_{ij} = 0, \tag{85}$$

and the source term of the entropy equation reduces to

$$TQ = \mathbf{X}^C \cdot \mathbf{Y}^C + \mathbf{X}^D \cdot \mathbf{Y}^D + \sum_{\alpha=1}^{\nu} \mathbf{X}^\alpha \cdot \mathbf{Y}^\alpha. \tag{86}$$

5.2 Vector currents and forces

System (82) now reduces to linear relations between vectors, the coefficients being second order tensors:

$$Y_i^C = L_{ik}^{CC} X_k^C + L_{ik}^{CD} X_k^D + \sum_{\beta=1}^{\nu} L_{ik}^{C\beta} X_k^\beta \tag{87}$$

$$Y_i^D = L_{ik}^{CD} X_k^C + L_{ik}^{DD} X_k^D + \sum_{\beta=1}^{\nu} L_{ik}^{D\beta} X_k^\beta \tag{88}$$

$$Y_i^\alpha = L_{ik}^{C\alpha} X_k^C + L_{ik}^{D\alpha} X_k^D + \sum_{\beta=1}^{\nu} L_{ik}^{\alpha\beta} X_k^\beta, \quad (\alpha = 1 \dots \nu), \tag{89}$$

where Onsager symmetry is accounted for. Remembering that $Y_i^\alpha = I_k^\alpha$, Eq. (36) leads to three constraints on the L-coefficients of Eq. (89):

$$\sum_{\alpha=1}^{\nu} L_{ik}^{C\alpha} = \sum_{\alpha=1}^{\nu} L_{ik}^{D\alpha} = \sum_{\alpha=1}^{\nu} L_{ik}^{\alpha\beta} = 0. \tag{90}$$

The linear combinations above show the possibilities of constructing models where interactions between thermal conduction, fluid flow, and mass diffusion are quantified by choosing the L coefficients. Also, thermal conduction and permeability can be modelled by second order tensors, through tensors L_{ik}^{CC} and L_{ik}^{DD}.

However, if one limits oneself to the minimal model where: (i) the diffusive entropy current, K_i/T, is only due to the $\partial_k T$ force, (ii) the $\mathfrak{F}_k^{\mathrm{F}}$ frictional force is only due to fluid velocity, V_i, (iii) each non-convective mass current, I_i^α, is only due to the gradients of the A^β, then one requires

$$L_{ik}^{CD} = 0, \quad \text{and} \quad L_{ik}^{C\alpha} = L_{ik}^{D\alpha} = 0 \quad \text{for all} \quad \alpha, \tag{91}$$

not violating the constraints in Eqs. (90).

Note that it is not possible to simplify the model to the extent that *all* cross-couplings are eliminated, since that would imply that $L_{ik}^{\alpha\beta}$ are zero except when the superscripts are equal: such a matrix would not obey the third constraint in (90). One could of course set $L_{ik}^{\alpha\beta} = 0$ for all α and β, but that would eliminate all the non-convective mass currents from the model, and probably make it useless.

In the minimal model one can add a fourth requirement to the three above: the upscaled fluid and the upscaled medium are isotropic. Then further simplifications result since the remaining second order tensors, L_{ik}^{CC}, L_{ik}^{DD}, and $L_{ik}^{\alpha\beta}$, that are properties of the solid and the fluid, must be invariant under rotations of the coordinate axes. Such tensors are called isotropic and it can be shown that an isotropic second order tensor is proportional to the Kronecker delta (see Ref. [15]). Isotropy thus introduces the following restrictions:

$$L_{ik}^{CC} = \ell^{CC}\delta_{ik}, \qquad L_{ik}^{DD} = \ell^{DD}\delta_{ik}, \qquad L_{ik}^{\alpha\beta} = \ell^{\alpha\beta}\delta_{ik}, \tag{92}$$

and one gets

$$\mathbf{Y}^C = \ell^{CC}\mathbf{X}^C, \qquad \mathbf{X}^D = \left(1/\ell^{DD}\right)\mathbf{Y}^D, \qquad \mathbf{Y}^\alpha = \sum_{\beta=1}^{\nu} \ell^{\alpha\beta}\mathbf{X}^\beta. \tag{93}$$

The source term of the entropy equation is now

$$TQ = \ell^{CC}\left|\mathbf{X}^C\right|^2 + \left(1/\ell^{DD}\right)\left|\mathbf{Y}^D\right|^2 + \sum_{\alpha=1}^{\nu}\sum_{\beta=1}^{\nu} \ell^{\alpha\beta}\mathbf{X}^\alpha \cdot \mathbf{X}^\beta. \tag{94}$$

Keeping in mind that the $\ell^{\alpha\beta}$-matrix is symmetric, and that the sum of all elements on the same line or the same column is zero, it can be shown that

$$\sum_{\alpha=1}^{\nu}\sum_{\beta=1}^{\nu} \ell^{\alpha\beta}\mathbf{X}^\alpha \cdot \mathbf{X}^\beta = -\sum_{\alpha,\beta,\alpha<\beta} \ell^{\alpha\beta}\left|\mathbf{X}^\alpha - \mathbf{X}^\beta\right|^2. \tag{95}$$

According to the expression for the source of entropy above, the remaining phenomenological coefficients must then satisfy

$$\ell^{CC} > 0, \qquad \ell^{DD} > 0, \qquad \ell^{\alpha\beta} < 0, \quad (\alpha < \beta). \tag{96}$$

These coefficients are determined below.

Using the notation of expressions (80) in the first of expressions (93) one gets K_i proportional to $\partial_i T$. Now K_i/T is an entropy transport by diffusion, so that K_i is heat transport by diffusion. To recover Fourier's law one sets

$$\ell^{CC} = k/T, \quad \text{implying} \quad K_i = -k\partial_i T, \tag{97}$$

where $k > 0$ is the thermal conductivity of the averaged solid–fluid system. See Ref. [3].

The second of expressions (93) gives $\mathfrak{F}_i^{\mathrm{F}} = V_i/\ell^{DD}$, and it is shown below that the choice

$$\ell^{DD} = \frac{K}{\Phi^2 \eta}, \quad \text{implies} \quad V_i = \frac{K}{\Phi\eta}\left(\partial_i P^{\mathrm{bF}} - Rf_i\right), \tag{98}$$

away from interphase regions. In this expression, η is the viscosity of the pore fluid, and K is the absolute permeability. As already mentioned, the phenomenological coefficients can be functions of the component densities and of the temperature, and it is known that η is such a function (see Section 3.4 in Ref. [3]). The possibility of letting K be such a function is not used in the minimal model. To prove the implication (98), one uses Eq. (42), with the upscaled stress tensor given by Eqs. (60), (61), and (85). Using Eq. (41) one obtains:

$$V_i = -\frac{K}{\Phi\eta}\left[\partial_i P^{\mathrm{bF}} - Rf_i + R(\partial_t V_i + V_k \partial_k V_i) - \sum_{\alpha=1}^{\nu} \Lambda^\alpha R^\alpha \partial_i \nabla^2 R^\alpha\right]. \tag{99}$$

The sum over α vanishes in the bulk fluid and one is left with the Darcy formula with an additional term, proportional to the material derivative of the velocity. One can carry out order of magnitude estimates of the three first terms in the square brackets above, in the manner of Section 4.1 of Ref. [3]. Using the numerical values given in the Appendix of the same reference, one easily finds that, if the gradient of pressure is of order 1, then the gravity term Rf_i is of order 10^{-3}, and the material derivative term is of order 10^{-9}. Neglecting the material derivative, one obtains the Darcy formula, modified by terms that only become significant inside the interphase regions. Specialised to a one component fluid, this modified Darcy formula is:

$$V_i = -\frac{K}{\Phi\eta}\left[\partial_i P^{\mathrm{bF}} - Rf_i\right] + \Lambda \frac{KR}{\Phi\eta} \partial_i \nabla^2 R. \tag{100}$$

It is shown in Ref. [4] that the added non-Darcy term can, in some well-defined flow types, produce a relative permeability when its numerical contribution is taken away as an added term, then put back as a multiplicative factor to the Darcy term. However, it is concluded in Ref. [4] that relative permeabilities cannot capture the full complexity of two-phase flow.

Given below is another version of V_i, that follows by application of the Gibbs-Duhem equation, obtained by differentiating $\mathcal{F}^{\mathrm{bF}}$ and using the expression of $\mathrm{d}\mathcal{F}^{\mathrm{bF}}$ (Eqs. (2)):

$$V_i = -\frac{K}{\Phi\eta}\left[\sum_{\alpha=1}^{\nu} R^\alpha \partial_i A^\beta + \mathcal{S}^{\mathrm{F}} \partial_i T\right], \tag{101}$$

where A^α is given by Eq. (70). Using now the notation of expressions (79) in the third of expressions (93) one gets the diffuse mass current of component α:

$$I_k^\alpha = -\Phi \sum_{\beta=1}^{\nu} \ell^{\alpha\beta} \partial_k A^\beta, \tag{102}$$

where it is reminded that the $\ell^{\alpha\beta}$-matrix is symmetric, its off diagonal elements are negative, and the elements of any line (or column) sum to zero. In addition, the elements can be functions of the densities. This last turns out to be extremely useful because the $\partial_k C^{F\beta}$ behave as $1/R^\beta$ when R^β goes to zero, which is not acceptable when the differential equations of the model are solved numerically. In the minimal model one then sets

$$\ell^{\alpha\beta} = -a\left[R^\alpha R^\beta\right]^2, \qquad (\alpha \neq \beta), \tag{103}$$

using the same positive number a for all elements. Each diagonal element $\ell^{\alpha\alpha}$ is then the negative sum of the off diagonal elements of line α. For an example of such a matrix in the case $\nu = 3$, see Refs. [2, 3].

5.3 Closing details

It is easy to see that the first two restrictions displayed in Eqs. (92) do not implicate any other assumptions done in the minimal model so that the restrictions can be lifted, either singly or together, thus allowing thermal conductivity and/or permeability to be represented by a second order tensor when experiments indicate that such upgrading is required.

A non-thermal version of the model consists of ν mass balance equations, see Eqs. (41), where the Darcy velocity is given by Eq. (101) and the mass diffusion velocities by Eqs. (102) and (103), the auxiliary variables being defined by Eqs. (70). Fot the thermal version of the model one needs to re-write the entropy equation, Eq. (75), in terms of temperature: see Section 2.3 of Ref. [3].

Start and boundary conditions must be supplied for the numerical solutions of the differential equations of the model. Special attention must be taken with the boundary conditions since the equations are of the fourth degree in the space variables. See Ref. [3] for a detailed presentation.

The transport coefficients, k and η, of Eqs. (97) and (98), are needed as functions of T and of the R^α. See Ref. [3] and references given there.

The central thermodynamical function of the model is the Helmholtz function, especially in the bulk, introduced by Eq. (1). It is calculated from the equation of state of the mixture considered, which must be van der Waals or related (Redlich-Kwong, ...) so that, for temperatures less than the critical, regions of unstable fluid insure the existence of interphase regions; association terms must be included for the polar molecules of the mixture (see Ref. [16] and references given there).

It is shown in Ref. [16] that, independently of the equation of state that is chosen, the Helmholtz function contains the sum of the Helmholtz functions of the components, each considered as a gas where molecular interactions are neglected (ideal). Each ideal gas Helmholtz function contains a function of T that drops out under differentiation when T is assumed constant, but is important for the thermal model in accounting for the energy stored in the internal degrees of freedom of the molecules. The Helmholtz function of one mole of an ideal gas of component α is (see Refs. [2, 16] and references given there):

$$f^{\text{id.gas},\alpha} = R_G T \ln \frac{\mathcal{V}^\alpha(T)}{e v^\alpha}, \tag{104}$$

where v^α is the molar volume, R_G is the gas constant, $e = \exp(1)$, and $\mathcal{V}^\alpha(T)$ is a function of temperature with the dimension of a volume per mole. $\mathcal{V}^\alpha$ depends on the atomic masses, the principal moments of inertia of the molecule, ... It can be obtained from statistical physics, as shown in Ref. [16].

It is briefly shown below that $\mathcal{V}^\alpha$ can also be obtained from an expression of $f^{\text{id.gas},\alpha}$ in terms of the heat capacity of the component. Assuming $Pv^\alpha = R_G T$, one easily finds that, with T and v^α as independent variables, the molar internal energy u^α (the "id.gas" superscript is dropped for simplicity) is independent of v^α, and that one can write

$$\mathrm{d}u^\alpha = c_V^\alpha(T)\mathrm{d}T, \tag{105}$$

where c_V^α is the molar heat capacity of component α. The molar entropy, s^α, obeying $T\mathrm{d}s^\alpha = \mathrm{d}u^\alpha + P\mathrm{d}v^\alpha$, one obtains

$$\mathrm{d}s^\alpha = c_V^\alpha(T)\mathrm{d}T/T + R_G \mathrm{d}v^\alpha/v^\alpha. \tag{106}$$

Integrating the last two displayed equations, one obtains $f^{\text{id.gas},\alpha}$ as $u^\alpha - Ts^\alpha$:

$$f^{\text{id.gas},\alpha} = u_0^\alpha - Ts_0^\alpha - R_G T \ln \frac{v^\alpha}{v_0^\alpha} + \int_{T_0}^{T} \left(1 - \frac{T}{T'}\right) c_V^\alpha(T')\mathrm{d}T', \tag{107}$$

where u_0^α and s_0^α are the internal energy and the entropy at a reference state where volume, temperature, and pressure are v_0^α, T_0, and $P_0 = R_G T_0 / v_0^\alpha$. Equating the right-hand sides of Eqs. (104) and (107), one obtains an expression for $T \ln \mathcal{V}^\alpha$ in terms of the experimentally measurable function c_V^α. As stated in Ref. [16], it is in fact the derivative with respect to T of $T \ln \mathcal{V}^\alpha$ that is needed in the differential equations of the thermal model. Assuming $s_0^\alpha = 0$, one obtains

$$\frac{\mathrm{d}}{\mathrm{d}T}\left[R_G T \ln \mathcal{V}^\alpha(T)\right] = R_G \ln \frac{e R_G T_0}{P_0} - \int_{T_0}^{T} \frac{c_V^\alpha(T')}{T'} \mathrm{d}T'. \tag{108}$$

Concerning examples of numerical solutions of the equations of the minimal model, see Ref. [2] for phase segregation, and for coning at uniform temperature; see Ref. [3] for an injection-production situation at variable temperature.

A. Appendices

A.1 Appendix 1: Assumptions

A0: None of the chemical species completely wets the rock
A1: There are no sources or sinks
A2: There is no loss of energy by radiation
A3: There are no chemical reactions between the chemical components
A4: The solid is perfectly rigid
A5: The external force per unit volume f_i is approximately constant in the averaging volume

A6: At each point, the difference between the solid and fluid temperatures is negligible

A7: The Λ^α are constant numbers inside the porous medium

A8: The porosity Φ is uniform in space

A.2 Appendix 2: Derivatives of distibutions – Formulas

The following is a set of formulas for the space derivatives of distributions having discontinuities across a given surface Σ. For proofs, see [12, 13].

Consider a function $f(\mathbf{x},t)$ symbolising a physical quantity, continuous in time but dicontinuous in space across a surface Σ that divides space in two regions, called A-side and B-side; $f = f^B$ on the B-side, $f = f^A$ on the A-side. Σ has a normal vector $\mathbf{n}$ at each point, pointing towards A:

$$f(\mathbf{x},t) = \begin{cases} f^A(\mathbf{x},t) & \mathbf{x} \text{ on the A-side of } \Sigma \\ f^B(\mathbf{x},t) & \mathbf{x} \text{ on the B-side of } \Sigma, \end{cases} \tag{109}$$

One now defines the following regular distributions:

$$\{\partial_k f(\mathbf{x},t)\} = \begin{cases} \partial_k f^A(\mathbf{x},t) & \mathbf{x} \text{ on the A-side of } \Sigma \\ \partial_k f^B(\mathbf{x},t) & \mathbf{x} \text{ on the B-side of } \Sigma, \end{cases} \tag{110}$$

$$\{\partial_t f(\mathbf{x},t)\} = \begin{cases} \partial_t f^A(\mathbf{x},t) & \mathbf{x} \text{ on the A-side of } \Sigma \\ \partial_t f^B(\mathbf{x},t) & \mathbf{x} \text{ on the B-side of } \Sigma. \end{cases} \tag{111}$$

Then [12, 13].

$$\partial_t f = \{\partial_t f\}, \tag{112}$$

$$\partial_k f = \{\partial_k f\} + \left[\left(f^A\right)^\Sigma - \left(f^B\right)^\Sigma\right] n_k\,\delta_\Sigma, \tag{113}$$

where δ_Σ is a surface-Dirac distribution, its action $\langle \delta_\Sigma, \varphi \rangle$ on a so-called test function φ [13] being defined as:

$$\langle \delta_\Sigma, \varphi \rangle = \int_\Sigma \varphi \, \mathrm{d}\Sigma. \tag{114}$$

A.3 Appendix 3: Formulas concerning tensors

Let Z_{ij} be an arbitrary second order tensor. It can be shown (see Ref. [15]) that the following expression has general validity:

$$Z_{ij} = Z\delta_{ij} + \hat{Z}_{ij} + \varepsilon_{ijk}\tilde{Z}_k, \text{ where } \begin{cases} Z = \dfrac{1}{3} Z_{kk}, \\ \hat{Z}_{ij} = \dfrac{1}{2}\left(Z_{ij} + Z_{ji}\right) - \dfrac{1}{3} Z_{kk}\delta_{ij}, \\ \tilde{Z}_k = \dfrac{1}{2}\varepsilon_{klm} Z_{lm}. \end{cases} \tag{115}$$

If Z_{ij} is a tensor, then Z is a scalar, $\hat{Z}$ is a symmetric traceless tensor, and $\tilde{Z}_k$ is a pseudo-vector. A most useful property is as follows: if U_{ij} and V_{ij} are two tensors then

$$U_{ij}V_{ij} = 3UV + \hat{U}_{ij}\hat{V}_{ij} + 2\tilde{U}_i\tilde{V}_i. \tag{116}$$

Author details

Paul Papatzacos
University of Stavanger, Stavanger, Norway

*Address all correspondence to: paul.papatzacos@lyse.net

References

[1] Papatzacos P: Macroscopic two-phase flow in porous media assuming the diffuse-interface model at pore level. Transport in Porous Media 2002;49: 139–174. DOI: 10.1023/A: 1016091821189

[2] Papatzacos P: A model for multiphase and multicomponent flow in porous media, built on the diffuse-interface assumption. Transport in Porous Media 2010;82:443–462. DOI: 10.1007/ s11242-009-9405-2

[3] Papatzacos P: A model for multiphase, multicomponent, and thermal flow in neutrally wetting porous media, built on the diffuse-interface assumption. Journal of Petroleum Science and Engineering 2016;143:141–157. DOI: 10.1016/j. petrol.2016.02.027

[4] Papatzacos P, Skjæveland SM: Diffuse-interface modelling of two-phase flow for a one-component fluid in a porous medium. Transport in Porous Media 2006;65:213–236. DOI: 10.1007/ s11242-005-6081-8

[5] van der Waals JD: The thermodynamic theory of capillarity under the hypothesis of a continuous variation of density, Journal of Statistical Physics 1979;20:200–244. DOI: 10.1007/BF01011514

[6] Korteweg DJ: Sur la forme que prennent les equations etc, Archives Neerlandaises des Sciences Exactes et Naturelles, 1901;II:6–20. DOI: Not known

[7] Anderson DM, McFadden GB, Wheeler AA: Diffuse-interface methods in fluid mechanics. Annual Review of Fluid Mechanics 1998;30:139–165. DOI: 10.1146/annurev.fluid.30.1.139

[8] Ganesan V, Brenner H: A diffuse-interface model of two-phase flow in porous media. Proceedings of the Royal Society London A 2000;456:731–803. DOI: 10.1098/rspa.2000.0537

[9] Adler PM, Brenner H: Multiphase phase flow in porous media. Annual Review of Fluid Mechanics 1988;20:35–59. DOI: 10.1146/annurev. fl.20.010188.000343

[10] Papatzacos P, Skjæveland SM: Relative permeability from thermodynamics. Society of Petroleum Engineers Journal 2004;9:213–236. DOI: 10.2118/87674-PA

[11] Hirschfelder JO, Curtiss CF, Bird RB: Molecular theory of gases and liquids, John Wiley and Sons, Inc. New York, Chapman and Hall, Limited, London, 1954. DOI: 10.1002/ pol.1955.120178311

[12] Marle CM: On macroscopic equations governing multiphase flow with diffusion and chemical reactions in porous media. International Journal of Engineering Science 1982;20:643–662. DOI: 10.1016/0020-7225(82)90118-5

[13] Appel W: Mathematics for physics and physisists, Princeton University Press 2007. ISBN-13: 978–0–691-13102-3

[14] Weinstock R: Calculus of Variations, Dover 1974. ISBN-13: 0–486–63069-2

[15] Reichl LE: A Modern Course in Statistical Physics, 2nd Edition, John Wiley and Sons 1998. ISBN: 0–471–59520-9

[16] Papatzacos P. The Helmholtz free energy of pure fluid substances and fluid mixtures. In: Proceedings of the 8th International Conference on Heat Transfer, Fluid Mechanics and Thermodynamics, (HEFAT2011); 11–13 July 2011; Mauritius.

Chapter 2

Multiscale Modeling of Non-Isothermal Fluid Transport Involved in Drying Process of Porous Media

Kieu Hiep Le

Abstract

To preserve the product quality as well as to reduce the logistics and storage cost, drying process is widely applied in the processing of porous material. In consideration of transport phenomena that involve a porous medium during drying, the complex morphology of the medium, and its influences on the distribution, flow, displacement of multiphase fluids are encountered. In this chapter, the recent advanced mass and energy transport models of drying processes are summarized. These models which were developed based on both pore- and continuum-scales, may provide a better fundamental understanding of non-isothermal liquid–vapor transport at both the continuum scale and the pore scale, and to pave the way for designing, operating, and optimizing drying and relevant industrial processes.

Keywords: drying model, multiscale – modeling, porous media, pore-network model, continuum model, diffusion model, upscaling strategy

1. Introduction

In both natural systems (i.e. clays, aquifers, oil and gas reservoirs, plants, and biological tissues) and industrial systems (i.e. fuel cells, concrete, textiles, polymer composites, capillary heat pipe, and paper, etc.), porous media are often encountered. From a morphological point of view, porous media is composed of a persistent solid matrix and an interconnected void space that can be occupied by fluid phases. In several porous systems, the void space is initially filled by liquid water. To maintain the product quality, prolong the storage time as well as reduce the logistics cost, the liquid water is needed to be removed by using various drying techniques. During the drying process, the complex morphology of the media and its influences on the capillary flow, liquid – vapor phase change under thermal effects are faced. These processes are complex by themselves [1, 2]. Additionally, it should be noted that the industrial dryers consume approximately 12% of the total energy used in manufacturing processes [3]. For several industrial sectors (i.e. tissue, food, and agriculture) the energy consumption ratio of drying may reach 33%. As a complicated, important, and high energy consumption process, the dryer designing, and operation should not be done by trial and error. A better fundamental understanding of this drying process at the pore- and continuous scales (i.e. understanding the influence of porous structure on the drying behavior, drying time determination, and energy consumption prediction) may help to design better products and processes.

The drying process of porous media may be considered at different scales. At different scales, the approaches used to describe the transport processes may be quite different [4]. For instance, in a drying plant, thousands or millions of drying products can be dried simultaneously and the state of the drying agent evolves over time and space. Thus, the intra-sample distribution of process parameters should be lumped to avoid the extremely expensive computational cost. As a result, the lumped model is the suitable model that should be implemented in drying plant simulation. Coming to the sample drying process, the internal transport process is taken into account. The heat, mass, and momentum conservation equations are developed based on the first physics principles. In a control volume, the process properties are assumed to be homogenous and isotropic and the impact of micro-heterogeneity on the transport process is omitted. The macro-scale models often result in a system of partial differential equations. Both globe and local drying behavior of the sample can be well predicted by using the continuous models if the size of the domain is large enough. For a thin layer of the porous medium, where micro-heterogeneity plays a role, the continuous models are not valid. Due to the interaction between the solid skeleton and the fluid phases, the transport equations should be derived for individual pores. Since the transport phenomena are directly investigated at the pore-scale, the impact of changes at the pore or microstructure such as wettability, pore shape, pore size distribution, and pore structure on drying behavior has been effectively interpreted by pore-scale models [5, 6]. One should notes that these models are not standing individually. The results of pore scale model can be upscaled to generate the effective parameters of macro scale models which can be reduced to the lumped models. An example multiscale modeling approach for wood particle drying process is presented in **Figure 1**.

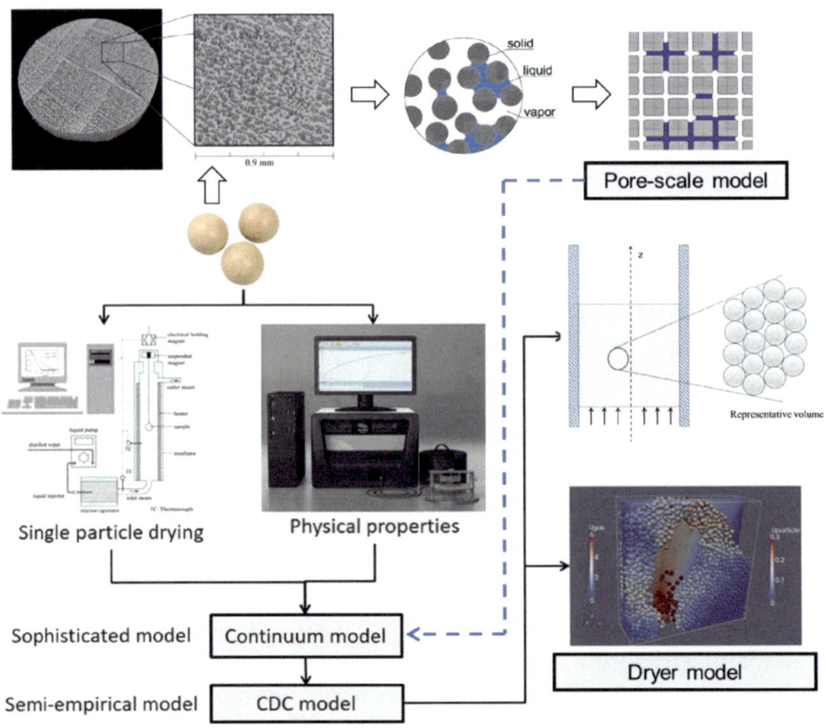

Figure 1.
Multiscale modeling approach for porous particle drying process.

In this chapter, a review of the recent advanced model in a selection of drying phenomena involved in porous media is illustrated. We discussed the state-of-the-art numerical methods as complementary ways to get more insight. The future challenges and the hint at solutions to accommodate for them are also given.

2. Pore-scale models

In 1950s, a new modeling approach to describe the transport process in the porous medium was proposed by Fatt. In his model, the void space of the porous media was modeled by a network of cylindrical tubes. In individual tubes, the mass transfer equation has been written and the discrete mass conservation equations are written for each pore node. This model has been named as pore network model. Afterward, several pore-scale models, such as the direct numerical approach, the volume of fluid approach, and method, have been presented in literature based on the idea of pore network model. Since the transport phenomena are considered directly at the pore-scale, the better fundamental understanding on the interaction between solid structure and fluid flow can be provided. As discussed in Section 1, the mentioned interaction can be upscaled to determine the macroscopic transport coefficients of the continuous models such as the relative permeabilities of the liquid–gas mixture, the liquid diffusivity.

In this section, we focus on the non – isothermal drying pore network models and their application in estimating the macroscopic transport coefficients.

2.1 Pore network model for superheated steam drying

Based on the invasion percolation algorithm, several non-isothermal pore network models were developed to simulate the hot air drying process [2, 7–9]. These models accounted capillary, gravity effects, viscous effects, and the transport of vapor by diffusion in the gas phase under the thermal effect. Recently, the pore network model has been applied to describe the non-isothermal mixture vapor - liquid water transport during the superheated steam drying process [10]. The features of this pore network model is the fully treatment of condensed liquid by introducing the newly formulated liquid invasion rules in a two-dimensional domain. In this chapter, this two-dimensional pore network model is extended to three-dimensional model in this work to simulate the transport processes inside a capillary porous medium undergoing superheated steam drying with two different heating modes: (i) the convective heating at the top surface (convective- heating mode) and (ii) the simultaneous convective and conductive heating at the top and bottom surfaces (contact-heating mode) (c.f. **Figure 2**).

Due to the absence of diffusion in the gas phase, the water vapor is transported by convective mechanism only and the Hagen-Poiseuille law is applied to calculate the vapor mass flow rate between two adjacent pores [10].

$$\dot{M}_{v,ij} = \frac{\pi r_{ij}^4}{8 \upsilon_v L}\left(p_i - p_j\right) \tag{1}$$

where the vapor mass flow rate in the cylindrical throat connecting pores i and j is denoted by $\dot{M}_{v,ij}$, p_i and p_j present the pressure of the vapor phase at pores I and j, respectively. The length and radius of throat are respectively indicated by L and r. The kinematic viscosity of the vapor phase is denoted by υ_v. Under the quasi-steady assumption, the mass conservation of water vapor in a empty/partially pore is written as

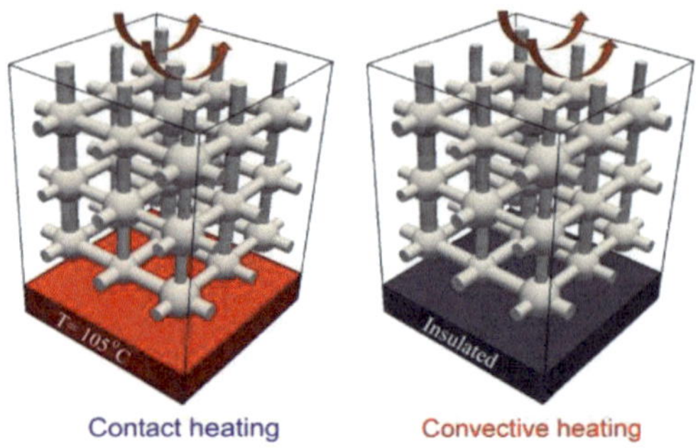

Figure 2.
Two different heating modes used in the pore network simulations.

$$\sum_j \dot{M}_{v,ij} - \sum_j \pi r_{t,ij}{}^2 \dot{m}_{\mathrm{evp/con},ji} - \pi r_{p,i}{}^2 \dot{m}_{evp/con,i} = 0 \tag{2}$$

Due to the non-uniform pore size distribution, the liquid within the network is transported under capillary action. Thus, the large pores/throats are preferentially emptied or filled to maintain the lowest liquid pressure according to Eq. 4. As a result, the liquid phase disintegrates into several liquid clusters. Since the liquid mass balance must be satisfied in the entire network as well as for each liquid cluster, the morphological information of liquid cluster is required. The Hoshen–Kopelman algorithm is applied for labeling the liquid clusters.

$$p_l = p_v - \frac{2\sigma(T)\cos\theta}{r_{p/t}} \tag{3}$$

To fully model the phase transition including both evaporation and condensation, in addition the emptying and refilling events, the liquid invasion represented in **Figure 3** is introduced in the present model. The time steps for the emptying and refilling events of meniscus pores or throats are computed by

$$\Delta t_{empty,cn/st,p/t} = \frac{V_{p/t}\rho_l(T)S_{p/t}}{\sum_{cn/st}\dot{M}_{evp/con}} \text{ and } \Delta t_{\mathrm{refill},cn/st,p/t} = -\frac{V_{p/t}\rho_l(T)\left(1 - S_{p/t}\right)}{\sum_{cn/st}\dot{M}_{evp/con}} \tag{4}$$

The time step for the first two liquid invasion events is calculated by

$$\Delta t_{in,cn/st} = -\frac{V_{p/t,}\rho_l(T)}{\sum_{cn/st}\dot{M}_{evp/con}} \tag{5}$$

The pore/throat saturations after the capillary re-equilibration event are computed by

$$S_{t,ih} = 1 \text{and} S_{p,h} = \frac{V_{p,h} - V_{t,ih}}{V_{p,h}} \tag{6}$$

The thermal energy supplied to a control volume (c.f. **Figure 4**) lead to a change of the enthalpy in the control volume: the control volume temperature increases, and the phase transition occurs. Based on a fully implicit scheme, the energy

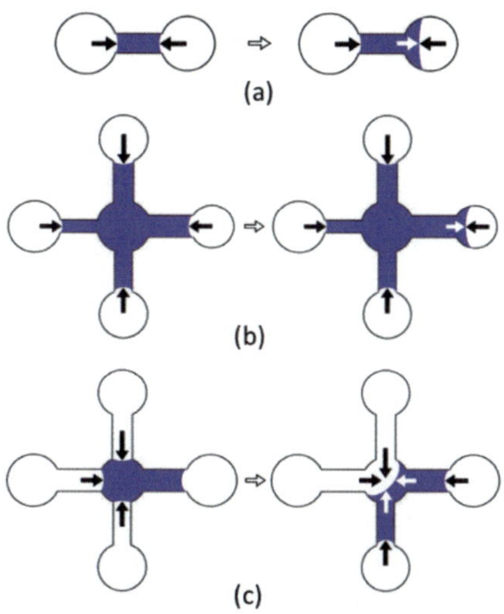

Figure 3.
The 2D sketch of liquid invasion rules: Liquid invasion from single liquid throat to its neighboring pore (a), liquid invasion from fully filled liquid throat in a liquid cluster to its neighboring empty pore (b), liquid redistribution from fully filled pore in a liquid cluster to its neighboring empty throat (c).

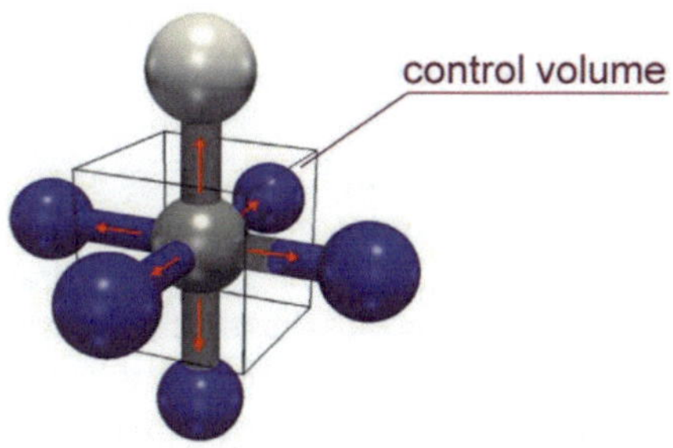

Figure 4
Heat balance at a vapor pore. The arrows indicate the direction of the heat flux towards each face of the control volume.

balance equation written for the control volume around pore i in time step Δt is discretized as

$$\sum_j -A_{cv,ij}\lambda_{eff,ij}\frac{T_{i,t+\Delta t}-T_{j,t+\Delta t}}{L}=\frac{\rho_{eff,i}c_{eff,i}V_i}{\Delta t}(T_{i,t+\Delta t}-T_i) + \Delta H_{evp}\sum_j \pi r_{t,ij}{}^2\dot{m}_{evp/con,ji}+\Delta H_{evp}\pi r_{p,i}{}^2\dot{m}_{evp/con,i} \tag{7}$$

The convective heat transfer boundary condition presented in Eq. 8 is applied for both heating modes. At the bottom of the network, perfect thermal insulation (Eq. 9) is imposed for the convective heating mode, whereas uniform constant temperature (Eq. 10) is set for the contact-heating mode.

$$\left[\lambda_{eff}\frac{\partial T}{\partial z}\bigg|_{y=L_y}+\Delta H_{evp}\rho_l v_l\big|_{y=L_y}\right]\bullet n=\alpha\left(T_{steam}-T|_{y=L_y}\right) \tag{8}$$

$$\frac{\partial T}{\partial y}\bigg|_{y=0}=0 \tag{9}$$

$$T|_{y=L_y} = T_{bottom} \tag{10}$$

The simulations are carried out for a 10 × 10 × 10 pore network with throat radius distribution 100 ± 10 μm, pore radius distribution 250 ± 10 μm and uniform distance between two adjacent pores L = 1000 μm. Because of the high computational demand required for a full simulation of drying, only one realization of network was considered here. The solid skeleton of the network is made of typical glass with thermal conductivity λ_s = 1 W/mK. The steam velocity at the network surface is 5 m/s. For the contact-heating mode, a uniform temperature boundary condition Tb_{ottom} = 105°C is imposed at the network bottom (c.f. Eq. 10), whereas the impermeable heat transfer boundary condition is set for the convective heating mode (c.f. Eq. 9). To depict the influence of heating mode on drying characteristics, the evolution of drying rate and of the local network saturation over network saturation obtained for these two heating modes are shown in **Figures 5** and **6**.

Since a large amount of thermal energy was supplied to the network with the contact-heating mode, the drying rate remains high and drying time is thus short compared to the network with the convective-heating mode. Also, since the surface temperature of the network with the contact-heating mode reaches boiling temperature fast, the surface condensation period is short and thus the amount of condensed liquid at the network surface is less, compared to the convective-heating mode. Moreover, it can be seen that the semi-constant drying rate period obtained from the network with the contact-heating mode is long compared to the convective heating model. This tendency can be explained by the temperature dependency of the surface tension; a lower surface tension obtained at high temperature leads to a lower capillary pressure and a higher liquid pressure at the liquid–vapor-solid interface. The pores/throats near the bottom of the network with the contact-heating mode are preferentially emptied. As a result, the vapor fingers appear markedly, and the bottom of the network is dried completely when the middle and top zones are still partly wet. The impact of vapor fingering on the phase distribution is shown in **Figure 5**. More number of liquid clusters is obtained with the contact-heating mode compared to the convective-heating mode. On the other hand, the network with the convective-heating mode is dried evenly from the top to the bottom zone. Since the top and middle zones remain saturated for long time, an extended quasi-constant drying rate period is obtained for the network with the contact-heating mode (c.f. **Figure 6**).

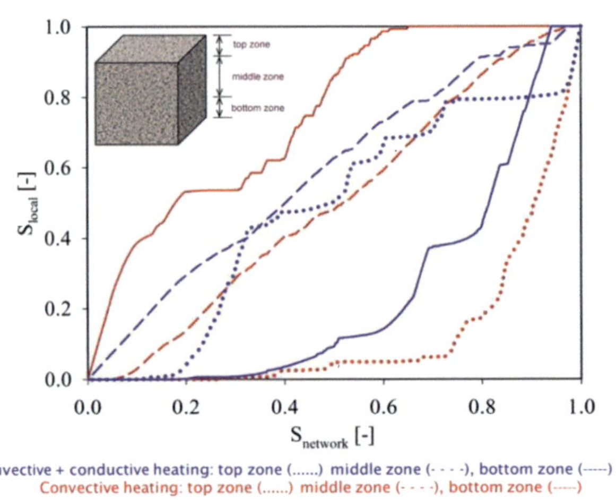

Figure 5.
Impact of heating mode on the local saturation evolution over network saturation.

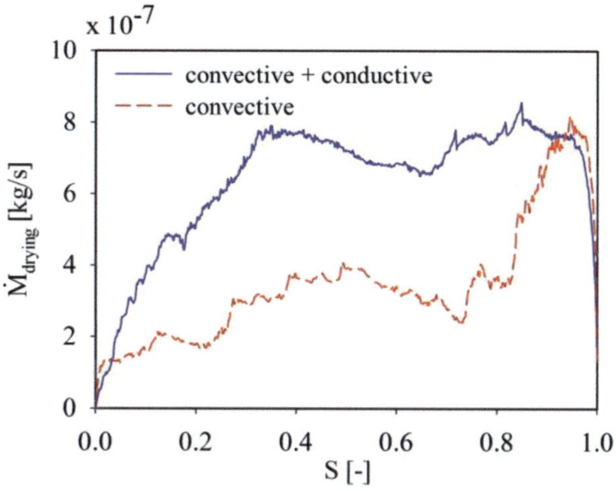

Figure 6.
Impact of heating mode on the drying kinetic curve.

2.2 Effective parameters of continuous model assessed from the pore network modeling

As mentioned in the previous section, the heat and mass transfer inside the porous domain can be simulated by using pore network models. However, since the pore network model considers the non-isothermal fluid flow in the pore individually, the number of the discrete equations of the model increases with the size of the domain. Thus, the pore network model cannot be used to simulate the macroscale drying process of porous media which is consists of billions of pores. To describe the drying process of macroscopic porous material, the continuous models developed based on the macroscopic effective transport parameters should be used. These effective parameters are often correlated from a serial of experimental data. To provide a fundamental understanding of the link between the transport coefficient and the fluid transport mechanisms, the transport coefficients are revisited by using the pore network simulation results. For example, the influence of network saturation on the non-local equilibrium effect, the effective moisture transport is investigated in the works of Kharaghani et al. [11–13] for the isothermal drying process. The incorporation of these effective parameters in the highly non-isothermal drying process is still an open issue. In the future, the impact of non-isothermal pore scale fluid transport on the effective parameter should be taken into account.

3. Continuum-scale models

In continuum-scale models, the porous media is assumed isotropic and homogeneous. Both solid structure and fluid properties are averaged in a representative elementary volume (REV) based on the volume averaging technique. The size of REV should not large enough to avoid the fluctuations of the morphological properties due to micro heterogeneity [14, 15]. The energy and mass conservation equations are derived based on the first physical principles using effective parameters.

3.1 Diffusion model

In diffusion model, the mixture of liquid water and vapor water flow is considered as single phase named moisture and the gradient of moisture content is solely driving force of moisture transport. Since the water diffusion flow leads to the evolution of water concentration in the control volume, the mass conservation of water is derived as

$$\frac{\partial \rho_0 X}{\partial t} + \frac{\partial}{\partial x} \cdot \left[-D_{eff} \frac{\partial}{\partial x} (\rho_0 X) \right] = 0. \tag{11}$$

In Eq. 11, the apparent density of the dry porous medium where the void volume is taken into account is computed as $\rho_0 = \frac{M_s}{V}$ (kg dry solid/m^3), $X = \frac{M_l}{M_s}$ (kg water/kg dry solid) is the dry-based moisture content of the wet solid. D_{eff} (m^2/s) is the effective diffusivity of moisture in the porous medium. Similarly, the energy conservation equation is written as the change of energy density caused by the enthalpy flow which consists of contributions from diffusive heat flow and heat conduction, namely,

$$\frac{\partial}{\partial t} \left[\left(\rho_0 c_{p,s} + \rho_0 c_{p,l} X \right) T \right] - \nabla \cdot \left[D_{eff} c_{p,l} T \nabla (\rho_0 X) \right] - \nabla \cdot \left[\lambda_{eff} \nabla (T) \right] = 0. \tag{12}$$

In Eq. 12, $c_{p,s}$ and $c_{p,l}$ (J/kg.K) denote the specific heat capacity of solid and liquid water, respectively; λ_{eff} (W/m.K) denotes the effective thermal conductivity of the porous medium.

To perform the diffusional model simulation, the effective thermal conductivity and effective moisture diffusivity needed to be known. Several methods such as fitch method, laser flash, hot-disk method and hot-wire method, can be used to measure the effective thermal conductivity directly whereas measurement of effective moisture diffusivity is rather more complex. Generally, the effective diffusivity of wetted porous materials often decreases during the water dehydration process since the liquid water is strongly captured in the small pores under a higher capillary action compared to large pores [16, 17]. Recently, Khan et al. [18] reported that the moisture diffusivity for isotropic porous materials, therein food products, can be described as a function of moisture content as

$$D_{eff} = D_{ref} \left(\frac{1 + X_0}{1 + X} \frac{\rho_l (1 + X) + \rho_0 X}{\rho_l (1 + X_0) + \rho_0 X_0} \right)^2, \tag{13}$$

where D_{ref} (m^2/s) is the reference diffusivity, X_0 is the initial moisture content. The question raised here is how to determine the reference effective diffusivity D_{ref}. Instead of measurement directly, it can be computed numerically by an optimization routine using invert method. The objective optimizing function $g(D_{eff})$ is the sum of square of difference between numerical moisture content $X_{i,sim}$ and experimental $X_{i,\exp}$ as

$$g(D_{eff}) = \sum_{i=1:end} \left(X_{\exp}(t_i) - X_{sim}(t_i) \right)^2, \tag{14}$$

where t_{i} denotes the time interval i of the data sampling. By minimizing the value of objective, the numerical moisture evolution is fitted to the experimental observation and the value of effective diffusivity is obtained. For an example, this method is applied for single wood particle drying process in superheated steam environment. The detail description of the geometrical information, initial and boundary conditions is presented in Le et al. [19]. The effective diffusivity of 6.633×10^{-9} m^2/s determined using the experimental data observed at a drying temperature of 140°C can be used reliable for a large range of drying temperature as shown in **Figure 7**. It implies that the moisture diffusivity model presented in Eq. 13 can be used resonable for wooden porous media.

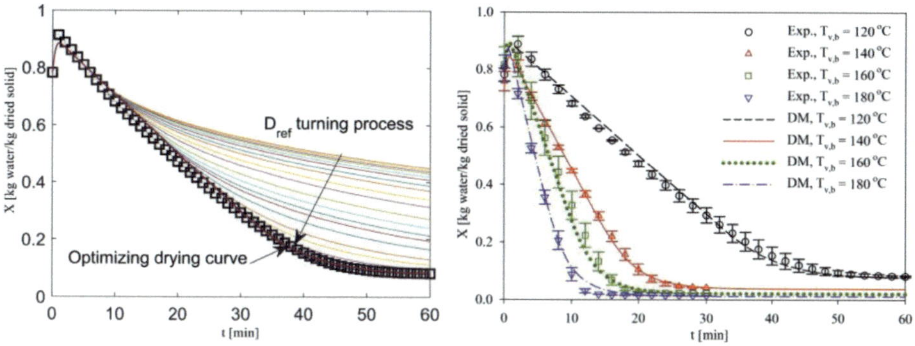

Figure 7.
The evolution of drying history during optimization process for D_{ref} determination (a) and experimental and numerical mean moisture content evolutions over time (b).

The validated diffusion model of wood drying is implemented in a CFD model is used to investigate the drying behavior of a pilot dryer. The sketch of dryer (L = 1000 mm, D = 3000 mm, and H = 500 mm) where 30 plates of wood (L = 900 mm, D = 200 mm, and H = 25 mm) is presented in **Figure 8**. The air with a temperature of 140°C and moisture content of 7 g ram vapor/kg dry air flows into the dryer with a velocity of 0.1 m/s. The initial moisture and temperature of wood are 0.61 kg water/kg dried solid and 20°C. The 2D velocity profile of drying agent is presented in **Figure 8** and the evolution of moisture content of plate number (1), (2), (3), and (4) are plotted together with the mean moisture content of all plates in **Figure 9**. The results indicate a noticeable moisture content maldistribution of wood plates which should be remedied by wood plate disturbing during the drying process.

3.2 Whitaker's continuum model

In the 1970s, the simultaneous heat, mass, and momentum transfer for non-isothermal drying process is proposed by Whitaker based on the volume averaging technique. In this model, all pore-level mechanisms for heat and mass transfer are considered. The liquid flow due to capillary forces, vapor and gas flow due to convection and diffusion are taken into account in this model. Using the volume averaging technique, the properties of fluid and solid phases such as velocity, density, pressure are averaged in REV. Afterward, the macroscopic differential equations were defined in terms of average field quantities. The detail descriptions of Whitaker's model can be found in Vu et al. [20, 21]. In this section, the mass and energy conservation equations are briefly recalled.

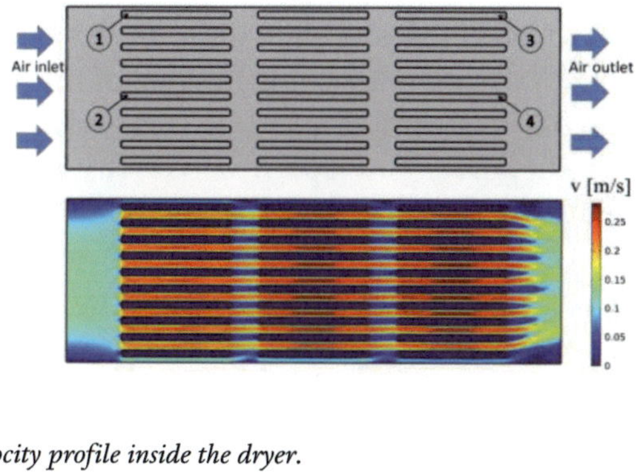

Figure 8.
Pilot dryer model and velocity profile inside the dryer.

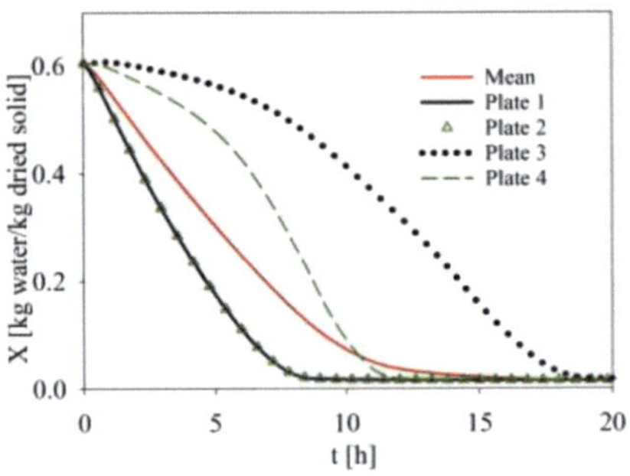

Figure 9.
Moisture content evolution over time of wood.

Mass conservation equation of liquid water

$$\frac{\partial}{\partial t}(\rho_l \varepsilon_l) + \nabla \cdot (\rho_l v_l) + \dot{m}_v = 0 \tag{15}$$

Mass conservation equation of water vapor

$$\frac{\partial}{\partial t}(\rho_v \varepsilon_g) + \nabla \cdot (\rho_v v_g) - \nabla \cdot \left(\rho_g D_{eff} \nabla \left(\frac{\rho_v}{\rho_g}\right)\right) - \dot{m}_v = 0 \tag{16}$$

Mass conservation of water in both vapor and liquid phases (sum of Eqs. 15 and 16)

$$\frac{\partial}{\partial t}(\rho_l \varepsilon_l + \rho_v \varepsilon_g) + \nabla \cdot (\rho_l v_l + \rho_v v_g) - \nabla \cdot \left(\rho_g D_{eff} \nabla \left(\frac{\rho_v}{\rho_g}\right)\right) = 0 \tag{17}$$

Mass conservation of air in the gas phase

$$\frac{\partial}{\partial t}(\rho_a \varepsilon_g) + \nabla \cdot (\rho_a v_g) - \nabla \cdot \left(\rho_g D_{eff} \nabla \left(\frac{\rho_a}{\rho_g}\right)\right) = 0 \tag{18}$$

Energy conservation equation

$$\begin{aligned} &\frac{\partial}{\partial t}(\rho_s \varepsilon_s h_s + \rho_l \varepsilon_l h_l + \rho_v \varepsilon_g h_v + \rho_a \varepsilon_g h_a) + \nabla \cdot [\rho_l h_l v_l + (\rho_v h_v + \rho_a h_a) v_g] \\ &-\nabla \cdot \left[h_a \rho_g D_{eff} \nabla \left(\frac{\rho_a}{\rho_g}\right)\right] - \nabla \cdot \left[h_v \rho_g D_{eff} \nabla \left(\frac{\rho_v}{\rho_g}\right)\right] - \nabla \cdot [\lambda_{eff} \nabla(T)] = 0 \end{aligned} \tag{19}$$

In these equations, ρ_l, ρ_v and ρ_s (kg/m3) are the mass density of liquid, vapor and air, respectively. $\varepsilon_l = \frac{V_l}{V}$, $\varepsilon_g = \frac{V_g}{V}$ are the volume fraction of the liquid and gas phases. The porosity of the medium is computed as $\psi = \varepsilon_l + \varepsilon_g$ and $\varepsilon_s = 1 - \psi$ denotes the volume fraction of the solid phase. The superficial velocity of the liquid and gas phased v_l and v_g (m/s) are calculated as

$$v_l = -\frac{KK_{r,l}}{\mu_l} \nabla p_l \text{ and } v_g = -\frac{KK_{r,g}}{\mu_g} \nabla p_g \tag{20}$$

where K (m^2) is the absolute permeability of the porous medium. $K_{r,l}$ and $K_{r,v}$ are the relative permeabilities of the liquid and gas phases, which are respectively calculated as $K_{r,l} = \left(\frac{\varepsilon_l}{\psi}\right)^2$ and $K_{r,g} = \left(1 - \frac{\varepsilon_l}{\psi}\right)^2$ [21, 22]. In Eq. 19, h_s, h_l, h_v and h_a (J/kg) are the specific enthalpy of solid, liquid, vapor and air, respectively. Using the assumption of constant specific heat capacity, the specific enthalpy of these components can be calculated respectively by $h_s = c_{p,s}(T - T_{ref})$, $h_l = c_{p,l}(T - T_{ref})$, $h_a = c_{p,a}(T - T_{ref})$ and $h_v = c_{p,v}(T - T_{ref}) + \Delta h_{evp}(T_{ref})$ at reference temperature T_{ref} = 0°C.

The Whitaker model's is suitable for the porous media made by impermeable solid phase. This model is applied to describe the drying process of wood material [23]. A good agreement between experimental and numerical data, shown in **Figure 10**, indicates the validity of the model.

For porous media structured by permeable solid, the liquid water is not only transported in the void space, but it also diffuses in the solid phase. In these hygroscopic porous materials, the liquid water is available in the form of both bound and free water:

$$M_l = M_{l,f} + M_{f,b} \tag{21}$$

where M_l, $M_{l,f}$ and $M_{f,b}$ denote the mass of total, free and bound liquid water, respectively. As a result, the total moisture content is the sum of the free and bound moisture content $X = X_{l,f} + X_{f,b}$. It should be noted that the bound water is mainly accumulated inside the cells of cellular porous products. A small amount of liquid water is located in small pores inside the solid matrix. In Le et al. [24], the bound water is assumed to be accumulated in the small pores of the macro-solid matrix. Thus, the bound water is removed by both local evaporation and liquid diffusion. The mass conservation equation of liquid can be rewrite as

$$\frac{\partial}{\partial t}\left(\varepsilon_{l,f}\rho_l + \overline{\rho}_b\right) + \nabla \cdot \left(\rho_l v_{l,f} + \overline{\rho_b v_b}\right) + \dot{m}_v = 0 \tag{22}$$

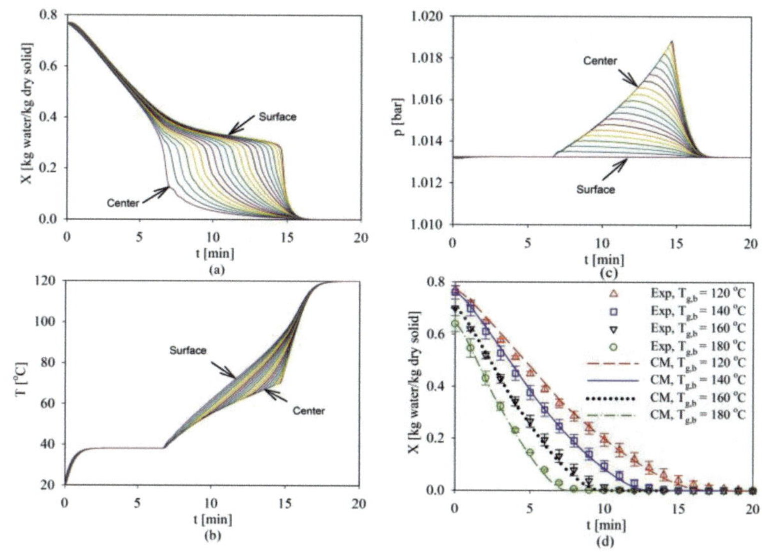

Figure 10.
The distribution of moisture content (a), temperature (b), vapor pressure (c) over drying time for wood particle drying at 120 °C and the comparison between experimental and numerical mean moisture content obtained with different drying temperature (d).

For the superheated steam drying process of rice seed, the bound water diffusivity can be calculated as

$$\overline{D}_b = \overline{D}_{b,ref}\left(\frac{\rho_w + \frac{X}{1+X}\rho_s}{\rho_w + \frac{X_0}{1+X_0}\rho_s}\right)^2 \text{with}\,\overline{D}_{b,ref} = 0.71 \times 10^{-9}\text{m}^2/\text{s} \qquad (23)$$

The internal structure of rice seed obtained by μ-CT and the comparison between the experimental observations and numerical results are presented in **Figure 11**. It indicates the validity of the proposed model.

Beside the porous material made by permeable solid phase, we consider the drying process of cellular plant product (i.e. fruits and vegetables) which is comprised of several types of cells such as parenchyma, collenchyma, and solute-conducting cells shown in **Figure 12** [25, 26]. In these materials, the water is mainly accumulated in the cell space. For several cellular product; i.e. eggplant, cucumber; the amount of intracellular water make up more than 95% of total water [27]. During the drying process, both intracellular and extracellular water is removed. The extracellular water is transported to the medium surface due to both capillary flow and internal evaporation. The changing of extracellular water content leads to the different of water potential between the cell space and intercellular void space which is the driving force of the intracellular water transport across the cell membrane as shown in **Figure 12**. A advance heat and mass transfer model for superheated steam drying process of cellular porous material was developed in Le et al. [28]. The conservation equations of extracellular and intracellular water are recalled as

$$\rho_l\frac{\partial \varepsilon_{l,ex}}{\partial t} + \rho_l\nabla.(v_{l,ex}) = -\dot{m}_v + j_{w,1}A_v\frac{\varepsilon_{l,ex}}{\varepsilon_{l,ex} + \varepsilon_{v,ex}} \qquad (24)$$

$$\frac{\partial(\rho_v\varepsilon_{v,ex})}{\partial t} + \nabla.(\rho_v v_{v,ex}) = \dot{m}_v + j_{w,2}A_v\frac{\varepsilon_{v,ex}}{\varepsilon_{l,ex} + \varepsilon_{v,ex}} \qquad (25)$$

$$\begin{aligned}\rho_l\frac{\partial \varepsilon_{l,ex}}{\partial t} + \rho_l\nabla.(v_{l,ex}) + \frac{\partial(\rho_v\varepsilon_{v,ex})}{\partial t} + \nabla.(\rho_v v_{v,ex})\\ -j_{w,1}A_v\frac{\varepsilon_{l,ex}}{\varepsilon_{l,ex} + \varepsilon_{v,ex}} - j_{w,2}A_v\frac{\varepsilon_{v,ex}}{\varepsilon_{l,ex} + \varepsilon_{v,ex}} = 0\end{aligned} \qquad (26)$$

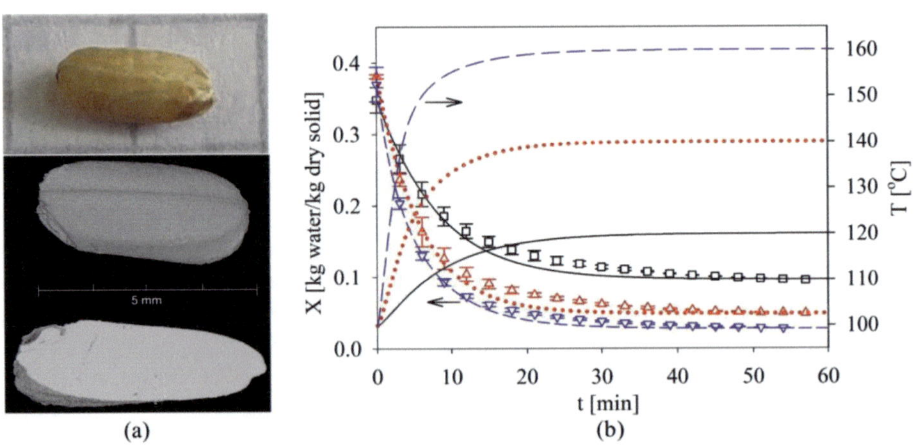

Figure 11.
The morphology of rice seed (a) and the numerical temperature and moisture content evolution over time obtained from diffusion simulations (b).

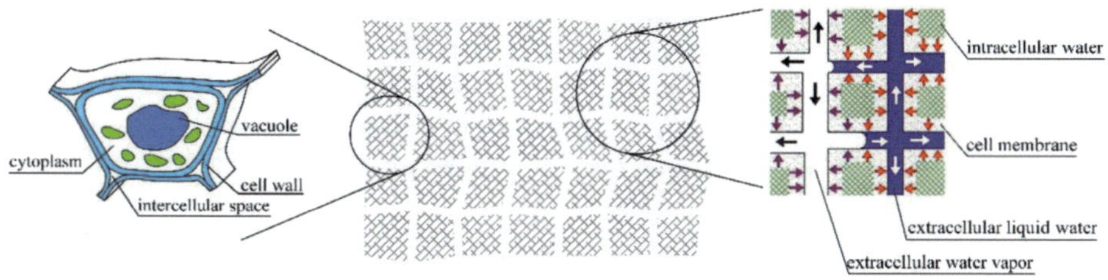

Figure 12.
Pictorial representation of extracelluar liquid water (in blue) and intracellular liquid water (in green) removal (reprinted from [28] with the permission).

In Eqs. 24–26, $\varepsilon_{l,ex} = \frac{V_{l,ex}}{V}$, $\varepsilon_{v,ex} = \frac{V_{v,ex}}{V}$ are the volume fractions of liquid water and water vapor in the intercellular void space, respectively. The volumetric evaporation flux is denoted by $\dot{m}_v (\text{kg/m}^3\text{s})$. A_v (m^2/m^3) is the volumetric specific area of the medium, which is the exchange area between cells and intercellular void space in a unit volume. The term $A_v \frac{\varepsilon_{l,ex}}{\varepsilon_{l,ex}+\varepsilon_{v,ex}}$refers to the cell surface area wetted by the liquid, where $\psi = \varepsilon_{l,ex} + \varepsilon_{v,ex}$ denotes the intercellular porosity of the medium. The diffusive water flux through the cell walls $j_{w,2}$ $(\text{kg/m}^2\text{s})$ serve as vapor sources. The mass conservation equation for the extracellular water vapor can thus be written as

$$\frac{\partial(\rho_v \varepsilon_{v,ex})}{\partial t} + \nabla.(\rho_v v_{v,ex}) = \dot{m}_v + j_{w,2} A_v \frac{\varepsilon_{v,ex}}{\varepsilon_{l,ex} + \varepsilon_{v,ex}} \tag{27}$$

After rearranging of Eqs. 24 and 25, the overall mass conservation equation for the water in the intercellular void space can be written as

$$\begin{aligned} \rho_l \frac{\partial \varepsilon_{l,ex}}{\partial t} + \rho_l \nabla.(v_{l,ex}) + \frac{\partial(\rho_v \varepsilon_{v,ex})}{\partial t} + \nabla.(\rho_v v_{v,ex}) \\ - j_{w,1} A_v \frac{\varepsilon_{l,ex}}{\varepsilon_{l,ex} + \varepsilon_{v,ex}} - j_{w,2} A_v \frac{\varepsilon_{v,ex}}{\varepsilon_{l,ex} + \varepsilon_{v,ex}} = 0 \end{aligned} \tag{28}$$

We assumed that the water diffusion across the cell membrane is the mechanism of cell water removal, the mass conservation equation for intracellular water inside the cell-matrix reads

$$\rho_l \frac{\partial \varepsilon_{l,\text{in}}}{\partial t} + j_{w,1} A_v \frac{\varepsilon_{l,ex}}{\varepsilon_{l,ex} + \varepsilon_{v,ex}} + j_{w,2} A_v \frac{\varepsilon_{v,ex}}{\varepsilon_{l,ex} + \varepsilon_{v,ex}} = 0 \tag{29}$$

where $\varepsilon_{l,in} = \frac{V_{l,in}}{V}$ denotes the volume fraction of cell water.

Using the local thermal equilibrium assumption, the heat balance equation of the control volume reads [26, 28].

$$\begin{aligned} \langle \rho c_p \rangle \frac{\partial T}{\partial t} + (c_{p,l} \rho_l v_{l,ex} + c_{p,v} \rho_v v_{v,ex}) \nabla \cdot T \\ = \nabla \cdot (\lambda_{eff} \nabla T) - \Delta h_{evp} \left(\dot{m}_v + j_{w,2} A_v \frac{\varepsilon_{v,ex}}{\varepsilon_{l,ex} + \varepsilon_{v,ex}} \right) \end{aligned} \tag{30}$$

In conservation equation system, the diffusive water fluxes through the cell membrane, i.e. $j_{w,1}$ and $j_{w,2}$ (red and violet arrows in **Figure 12**), need to be known. These diffusive fluxes are determined by using the concept of the water potential ϕ (Pa) and water conductivity of the cell membrane as

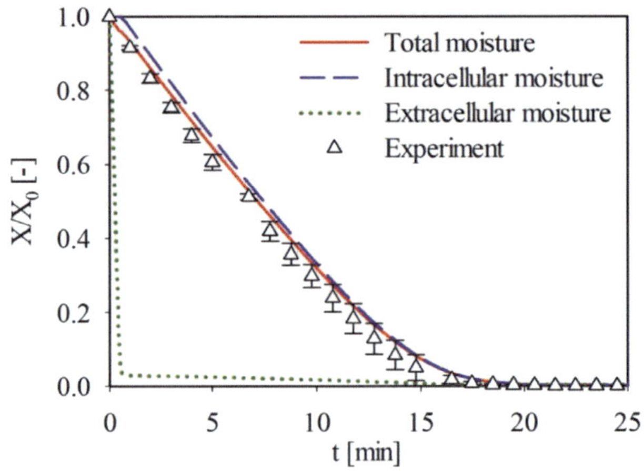

Figure 13.
Simulated and experimental evolution of the normalized total, intracellular and extracellular moisture content over time for potato drying in superheated steam environment at 180°C (replotted from [28] with the permission).

$$j_{w,1} = -\rho_l k_p (\phi_{w,in} - \phi_{l,ex}) \tag{31}$$

$$j_{w,2} = -\rho_l k_p (\phi_{w,in} - \phi_{v,ex}) \tag{32}$$

The water potential ϕ is computed as the difference of the total energy of one kilogram water molecules compared to liquid water molecule energy at the free liquid – water interface at 1 bar [25]. For the intracellular water, the water potential $\phi_{w,in}$, which was empirically determined in [29] as a function of the moisture content $\phi_{w,in} = f(X_{in})$. The potential of liquid water in the intercellular void space $\phi_{l,ex}$ is calculated as the capillary pressure, i.e. $\phi_{l,ex} = -p_c$. The water vapor potential $\phi_{v,ex}$ is computed from the gas theory based on the Gibbs free energy [30].

$$\phi_{v,ex} = \frac{\tilde{R} T \rho_l}{\tilde{M}_w} \ln \frac{p_{v,ex}}{p_{v,sat}} \tag{33}$$

This model has been used to describe the superheated steam drying process of potato. As can be seen in **Figure 13**, the experimental observations can be reflected fairly by numerical results. It implies the predictive potential of the proposed model. Furthermore, the extracellular moisture is removed very soon when the dry process commences. The drying process is mainly driven by the dehydration of intracellular water by the diffusion mechanism. Although the Whitaker's continuum approach is effectively used to predict the drying behavior, the parameters of the continuum model is seemingly difficult to be measured. Thus, in the future, with the help of pore-scale model, these effective parameters should be theoretically extracted from the pore level simulations.

4. Conclusions

In this chapter, some recent advance models on heat and mass transfer during non-isothermal drying process of porous media have been reviewed. It indicates that the drying process at pore-scale and macro-scale has been thoroughly investigated. However, a system study on the bridges between the pore-, macro- and plant scales shall be carried out. For example, several effective macroscopic parameters such as the capillary pressure curve or the relative permeabilities may be obtained from a post-processing of the Monte-Carlo simulation results of pore-scale modeling. The moisture diffusivity under the thermal effect also should be revisted and compared with the insothermal drying process. This upscaling statergy can help on both the product and system design, operation, and optimization.

Acknowledgements

This work is funded by the VIETNAMESE MINISTRY OF EDUCATION AND TRAINING (MOET) under research project B2021-BKA-12.

Author details

Kieu Hiep Le
School of Heat Engineering and Refrigeration, Hanoi University of Science and Technology, Hanoi, Vietnam

*Address all correspondence to: hiep.lekieu@hust.edu.vn

References

[1] M. Prat, Recent advances in pore-scale models for drying of porous media, Chemical Engineering Journal 86 (2002) 153–164.

[2] V.K. Surasani, T. Metzger, E. Tsotsas, Consideration of heat transfer in pore network modelling of convective drying, International journal of heat and mass transfer 51 (2008) 2506–2518.

[3] A.S. Mujumdar, Handbook of Industrial Drying, CRC Press, 2014.

[4] T. Defraeye, Advanced computational modelling for drying processes – A review, Applied Energy 131 (2014) 323–344.

[5] M. Sahimi, Flow and transport in porous media and fractured rock: From classical methods to modern approaches, 2nd rev. and enlarged ed. ed., Wiley-Vch Verlag GmbH & Co. KGaA, Weinheim, op. 2011.

[6] T. Sochi, Pore-Scale Modeling of Non-Newtonian Flow in Porous Media, PhD dissertation, London, 2007.

[7] S. Taslimi Taleghani, M. Dadvar, Two dimensional pore network modelling and simulation of non-isothermal drying by the inclusion of viscous effects, International Journal of Multiphase Flow 62 (2014) 37–44.

[8] V.K. Surasani, T. Metzger, E. Tsotsas, Drying Simulations of Various 3D Pore Structures by a Nonisothermal Pore Network Model, Drying Technology 28 (2010) 615–623.

[9] V.K. Surasani, T. Metzger, E. Tsotsas, Influence of heating mode on drying behavior of capillary porous media: Pore scale modeling, Chemical engineering science 63 (2008) 5218–5228.

[10] K.H. Le, A. Kharaghani, C. Kirsch, E. Tsotsas, Discrete pore network modeling of superheated steam drying, Drying Technology 35 (2017) 1584–1601.

[11] A. Attari Moghaddam, A. Kharaghani, E. Tsotsas, M. Prat, Kinematics in a slowly drying porous medium: Reconciliation of pore network simulations and continuum modeling, Physics of Fluids 29 (2017) 22102.

[12] X. Lu, A. Kharaghani, E. Tsotsas, Transport parameters of macroscopic continuum model determined from discrete pore network simulations of drying porous media: Throat-node vs. throat-pore configurations, Chemical engineering science 223 (2020) 115723.

[13] X. Lu, E. Tsotsas, A. Kharaghani, Insights into evaporation from the surface of capillary porous media gained by discrete pore network simulations, International journal of heat and mass transfer 168 (2021) 120877.

[14] P. Perré, The proper use of mass diffusion equations in drying modeling: Introducing the drying intensity number, Drying Technology 33 (2015) 1949–1962.

[15] S. Whitaker, Simultaneous heat, mass, and momentum transfer in porous media: A theory of drying, in: Advances in Heat Transfer Volume 13, Elsevier, 1977, pp. 119–203.

[16] E. Tsotsas, A.S. Mujumdar (Eds.), Modern Drying Technology, Wiley-Vch Verlag GmbH & Co. KGaA, Weinheim, Germany, 2007.

[17] P. Suvarnakuta, S. Devahastin, A.S. Mujumdar, A mathematical model for low-pressure superheated steam drying of a biomaterial, Chemical Engineering and Processing: Process Intensification 46 (2007) 675–683.

[18] M.I.H. Khan, C. Kumar, M.U.H. Joardder, M.A. Karim, Determination of

appropriate effective diffusivity for different food materials, Drying Technology 35 (2017) 335–346.

[19] K.H. Le, T.T.H. Tran, E. Tsotsas, A. Kharaghani, Superheated steam drying of single wood particles: Modeling and comparative study with hot air drying, Chem. Eng. Technol. 44 (2021) 114–123.

[20] H.T. Vu, E. Tsotsas, Mass and Heat Transport Models for Analysis of the Drying Process in Porous Media: A Review and Numerical Implementation, International Journal of Chemical Engineering 2018 (2018) 1–13.

[21] H.T. Vu, E. Tsotsas, A Framework and Numerical Solution of the Drying Process in Porous Media by Using a Continuous Model, International Journal of Chemical Engineering 2019 (2019) 1–16.

[22] K.H. Le, N. Hampel, A. Kharaghani, A. Bück, E. Tsotsas, Superheated steam drying of single wood particles: A characteristic drying curve model deduced from continuum model simulations and assessed by experiments, Drying Technology 36 (2018) 1866–1881.

[23] A. Kharaghani, K.H. Le, T.T.H. Tran, E. Tsotsas, Reaction engineering approach for modeling single wood particle drying at elevated air temperature, Chemical engineering science 199 (2019) 602–612.

[24] N. Hampel, K.H. Le, A. Kharaghani, E. Tsotsas, Continuous modeling of superheated steam drying of single rice grains, Drying Technology 37 (2019) 1583–1596.

[25] P.S. Nobel, Physicochemical and environmental plant physiology, 4th ed. ed., Academic Press, [Place of publication not identified], 2009.

[26] B. Xiao, J. Chang, X. Huang, X. Liu, A Moisture Transfer Model for Isothermal Drying of Plant Cellular Materials Based on the Pore Network Approach, Drying Technology 32 (2014) 1071–1081.

[27] A. Halder, A.K. Datta, R.M. Spanswick, Water transport in cellular tissues during thermal processing, AIChE J. 57 (2011) 2574–2588.

[28] K.H. Le, E. Tsotsas, A. Kharaghani, Continuum-scale modeling of superheated steam drying of cellular plant porous media, International journal of heat and mass transfer 124 (2018) 1033–1044.

[29] B. Xiao, J. Chang, X. Huang, X. Liu, A moisture transfer model for isothermal drying of plant cellular materials based on the pore network approach, Drying Technology 32 (2014) 1071–1081.

[30] P.S. Nobel, Physicochemical and environmental plant physiology, 4th ed., Academic Press, Amsterdam, Boston, 2009.

Section 2

Mathematical Models in Porous Media and Solutions

Chapter 3

Study on Approximate Analytical Method with Its Application Arising in Fluid Flow

Twinkle R. Singh

Abstract

This chapter is about the, Variational iteration method (VIM); Adomian decomposition method and its modification has been applied to solve nonlinear partial differential equation of imbibition phenomenon in oil recovery process. The important condition of counter-current imbibition phenomenon as $v_i = -v_n$, has been considered here main aim, here is to determine the saturation of injected fluid $S_i(x,t)$ during oil recovery process which is a function of distance ξ and time θ, therefore saturation S_i is chosen as a dependent variable while x $and t$ are chosen as independent variable. The solution of the phenomenon has been found by VIM, ADM and Laplace Adomian decomposition method (LADM). The effectiveness of our method is illustrated by different numerical.

Keywords: Variational Iteration method (VIM), Adomian decomposition method (ADM), Laplace Adomian decomposition method (LADM), nonlinear partial differential equations

1. Introduction

First, the variational iteration method was proposed by He [1] in 1998 and was successfully applied to autonomous ordinary differential equation, to nonlinear partial differential equations with variable coefficients. In recent times a good deal of attention has been devoted to the study of the method. The reliability of the method and the reduction in the size of the computational domain give this method a wide applicability. The VIM based on the use of restricted variations and correction functional which has found a wide application for the solution of nonlinear ordinary and partial differential equations, e.g., [2–10]. This method does not require the presence of small parameters in the differential equation, and provides the solution (or an approximation to it) as a sequence of iterates. The method does not require that the nonlinearities be differentiable with respect to the dependent variable and its derivatives and whereas the Adomian decomposition method was before the Nineteen Eighties, it was developed by Adomian [11, 12] for solving linear or nonlinear ordinary, partial and Delay differential equations. A large type of issues in mathematics, physics, engineering, biology, chemistry and other sciences have been solved using the ADM, as reported by many authors [13]. The Adomian decomposition method (ADM) [11–28] is well set systematic method for practical solution of linear or nonlinear and deterministic or stochastic operator equations, including ordinary differential equations (ODEs), partial differential equations

(PDEs), integral equations, integro-differential equations, etc. The ADM is considered as a powerful technique, which provides efficient algorithms for analytic approximate solutions and numeric simulations for real-world applications in the applied sciences and engineering. It allows us to solve both nonlinear initial value problems (IVPs) and boundary value problems (BVPs) [17, 29–46] without unphysical restrictive assumptions such as required by linearization, perturbation, ad hoc assumptions, guessing the initial term or a set of basic functions, and so forth. The accuracy of the analytic approximate solutions obtained can be verified by direct substitution. More advantages of the ADM over the variational iteration method is mentioned in Wazwaz [22, 28]. A key notion is the Adomian polynomials, which are tailored to the particular nonlinearity to solve nonlinear operator equations. A key concept of the Adomian decomposition series is that it is computationally advantageous rearrangement of the Banach-space analog of the Taylor expansion series about the initial solution component function, which permits solution by recursion. The selection behind choice of decomposition is nonunique, which provides a valuable advantage to the analyst, permitting the freedom to design modified recursion schemes for ease of computation in realistic systems.

Same way Laplace Adomian's Decomposition Method (LADM) was first introduced by Khuri [47, 48]. The Laplace Adomian Decomposition Method (LADM) is formed with combination of the Adomian Decomposition Method (ADM) Adomian [29, 49] and Laplace transforms. LADM is a promising method and has been applied in solving various nonlinear systems of differential equations [36, 50–56]. In a variety of applied sciences, systems of partial differential equations have attracted much attention e.g. [50, 57–75]. The general ideas and the essentiality of these systems are of wide applicability. Agadjanov [56] solved Duffing equation with the help of LDM. Elgazery [51, 76] had applied Laplace decomposition method for the solution of Falkner-Skan equation.

In the solution procedure of VIM; many repeated computations and computations of the unneeded forms, which take more time and effort beyond it, so a modification has been shown to reduce these unneeded forms.

On the other hand, few researchers have been discussed imbibition phenomenon in homogenous porous media with different point of view for example, researchers taking different perspectives for this phenomenon; [77, 78] and some others have analyzed it for homogeneous porous medium.

In this Present investigated model, Imbibition takes place over a small part of a large oil formatted region taken as a cylindrical piece of homogeneous porous medium. In this model, we have considered the important condition of counter-current imbibition phenomenon as $v_i = -v_n$, Our purpose is to determine the saturation of injected fluid $S_i(x,t)$ during oil recovery process which is a function of distance ξ and time θ, therefore saturation S_i has been chosen as a dependent variable while x and t are chosen as independent variable.

2. Imbibition phenomenon

It is the process by which a wetting fluid displaces a non-wetting fluid the initially saturates a porous sample, by capillary forces alone. Suppose a sample is completely saturated with a non-wetting fluid, and same wetting fluid is introduced on its surface. There will be spontaneous flow of wetting fluid into the medium, causing displacement of the non-wetting fluid. This is called imbibition phenomenon. The rate of imbibition is greater if the wettability of the porous medium, by the imbibed fluid, is higher.

The mathematical condition for imbibition phenomenon is given by Scheidegger [78]); viz,

$$v_n = -v_i$$

Where v_i & v_n are the seepage velocities of injected & native liquids respectively. The relation between relative permeability and phase-saturation,

$$k_i = S_i^3$$
$$k_n = 1 - \alpha S_n, \alpha = 1.11$$

Where k_i & k_n denotes fictitious relative permeability. S_i & S_n denotes saturations of injected and native liquids respectively.

3. Mathematical structure of the model

According to the Darcy's law, the basic equations of the phenomenon as; [78]

$$v_i = -\left(\frac{k_i}{\delta_i}\right)K\frac{\partial p_i}{\partial x} \tag{1}$$

$$v_n = -\left(\frac{k_n}{\delta_n}\right)K\frac{\partial p_n}{\partial x} \tag{2}$$

$$v_i = -v_n \tag{3}$$

$$p_c = p_n - p_i \tag{4}$$

$$\varphi\left(\frac{\partial S_i}{\partial t}\right) + \frac{\partial v_i}{\partial x} = 0 \tag{5}$$

$$\varphi\left(\frac{\partial S_n}{\partial t}\right) + \frac{\partial v_n}{\partial x} = 0 \tag{6}$$

Where v_i and v_n are the seepage velocities, k_i and kare the relative permeabilities δ_i and δ_n are the kinematic viscosities (which are constants), p_i and p_n are pressure of the injected and native liquid respectively, φ and K are the porosity and the permeability of the homogeneous porous medium; S_i is the saturation of the injected liquid; p_c is the capillary pressure and t is the time. The co-ordinate x is measured along the axis of the cylindrical medium, the origin being located at the imbibition face x=0.

Combing equations (1)-(5) and using the relation for capillary pressure as, $p_c = \beta S_i$ [70], we get,

$$\varphi\left(\frac{\partial S_i}{\partial t}\right) + \frac{\partial}{\partial x}\left[KD(S_i)\beta\left(\frac{\partial S_i}{\partial x}\right)\right] = 0 \tag{7}$$

Where $D(S_i) = \frac{k_i k_n}{\delta_n k_i + \delta_i k_n}$ and β being small capillary pressure coefficient.
It is assumed is that an average value of $D(S_i) = \overline{D}\ (S_i)$
Using the transformation,

$$\xi = \frac{x}{L}\ ,\theta = \frac{Lt}{\varphi L^2},\quad 0 \leq x \leq \frac{LS_{io}}{B}. \tag{8}$$

Eq. (7), becomes;

$$\left(\frac{\partial S_i}{\partial \theta}\right) + \beta\overline{D}(S_i)\frac{\partial^2 S_i}{\partial \xi^2} = 0$$

$\frac{\partial S_i}{\partial \theta} = -\beta\overline{D}(S_i)\frac{\partial {S_i}^2}{\partial \xi^2}$

$$\frac{\partial S_i}{\partial \theta} = \varepsilon\frac{\partial {S_i}^2}{\partial \xi^2} \text{ Where } \varepsilon = -\beta\overline{D}(S_i) \tag{9}$$

By the Hopf-Cole transformation [79, 80] equation (9) reduces to the Burger's equation.

$$S^*_{i_\theta} + S^*_i S^*_{i_\xi} = \varepsilon {S^*_i}_{\xi\xi} \tag{10}$$

With the condition
$S^*_i(\xi, 0) = S^*_{i_0} e^{\xi}$ at time $\theta = 0$ and $\xi > 0$

3.1 Solution of the Burger's equation by variational iteration method

To add the basic concepts of VIM, considering the below mentioned nonlinear partial differential equations:

$$\begin{aligned} &Lu(x,t) + Ru(x,t) + Nu(,t) = g(x,t), \\ &u(x,0) = e^x \end{aligned} \tag{11}$$

Where $L = \left(\frac{\partial}{\partial t}\right)$, R is a linear operator which has partial derivatives with respect to x, Nu(x,t) is a nonlinear term and g(x,t) is an inhomogeneous term.
As per the VIM [6, 7];

$$U_{n+1}(x,t) = U_n(x,t) + \int_0^t \lambda\{LU_n + \overline{RU_n} + \overline{NU_n} - g\}d\tau \tag{12}$$

Where λ is called a general Lagrange multiplier [81, 82] which can be identified optimally via vatiational theory, $\overline{RU_n}$ and $\overline{NU_n}$ are considered as restricted variations,
i.e. $\delta\overline{RU_n} = 0, \delta\overline{NU_n} = 0$ calculating variation with respect to U_n;

$$\begin{aligned} &\lambda'(\tau) = 0 \\ &1 + \lambda(\tau)_{\tau=t} = 0 \end{aligned} \tag{13}$$

The Lagrange multiplier, therefore, can be considered as λ=-1.
Now, substituting the multiplier in (12), then

$$U_{n+1}(x,t) = U_n - \int_0^t \{L(U_n) + R(U_n) + N(U_n) - g\}d\tau \tag{14}$$

$$S^*_{i_\theta} + S^*_i S^*_{i_\xi} = \varepsilon {S^*_i}_{\xi\xi} \tag{15}$$

With the constrain
$S^*_i(\xi, 0) = S^*_{i_0} e^{\xi}$ at time $\theta = 0$ and $\xi > 0$

To solve equation (10) by VIM, substituting in equation (14) by

$$RU_n = -U_{n_x}^2$$
$$NU_n = U_n(U_n)_{xx}$$

& g(x,t) = 0
And can obtain the following variational iteration formula:

$$S_{i\ n+1}^* = S_{i_n}^* - \int_0^\theta \left\{ S_{i_{n_\tau}}^* + S_{i_n}^* \left(S_{i_n}^* \right)_\xi - \varepsilon S_{i_{n\xi\xi}}^* \right\} d\tau \tag{16}$$

Using (14), the approximate solutions $U_n(x,t)$are obtained by substituting;

$$S_i^*(\xi, 0) = S_{i_0}^* e^\xi \tag{17}$$

Approximate solutions are given below;

$$S_{i_1}^* = S_{i_0}^* e^\xi - \beta_1^0 \theta; \quad where\ \beta_1^0 = \left(S_{i_0}^{*\,2} e^{2\xi} - \varepsilon S_{i_0}^* e^\xi \right)$$

$$S_{i_2}^* = S_{i_0}^* e^\xi - \beta_1^0 \theta + \beta_1^1 \frac{\theta^2}{2} \quad where\ \beta_1^1 = \beta_1^0 S_{i_0}^* e^\xi$$

Similarly,

$$S_{i_3}^* = S_{i_0}^* e^\xi - \beta_2^1 \frac{\theta^2}{2!} + \beta_2^2 \frac{\theta^3}{3!}$$

And so on … ..
Notes on VIM
From the analysis we can observed is this:

1. VIM can contain a series solution not exactly like ADM.

2. VIM needs many modifications to overcome the wasted time in the repeated calculations and unneeded terms.

To overcome these problems, following ADM and LADM is suggested.
Now applying ADM to equation (10); we get

$$S_i^*(\xi, \theta) = L_\theta^{-1} \left[S_i^* S_{i_\xi}^* - \varepsilon S_{i_{\xi\xi}}^* \right] \tag{18}$$

And recursive relation is:

$$S_i(\xi, 0) = e^\xi$$

Then:

$$S_{i_1}^*(\xi, \theta) = \beta_1^0 \theta$$

$$S_{i_2}^*(\xi, \theta) = \beta_{12}^0 \frac{\theta^3}{3} - \varepsilon \beta_2^1 \frac{\theta^2}{2}$$

$$S^*_{i_3}(\xi,\theta) = \beta^0_{123}\frac{\theta^4}{4} - \varepsilon\beta^1_3\frac{\theta^3}{3}$$

and so on …

Now, applying (LADM) Laplace transform with respect to t on both sides of (10);

$$S^*_i(x,t) = L^{-1}\left[\frac{1}{s}L\left[S^*_{i_0}S^*_{i_{0_\xi}} - \varepsilon S^*_{i_{0_{\xi\xi}}}\right]\right]$$

$$S^*_{i_1} = \beta^0_1 e^{\xi}\theta$$

$$S^*_{i_2} = \left(\beta^{0^2}_1 e^{2\xi} - \varepsilon\beta^0_1 e^{\xi}\right)\frac{\theta^2}{2!}$$

$$S^*_{i_3} = \left(\beta^3_0 - \varepsilon\beta^3_1\right)\frac{\theta^3}{3!}$$

And so on …

4. Interpretation

It is concluded that for the non linear partial differential equation of imbibitions phenomenon in oil recovery process, through graphs, it has been observed that the

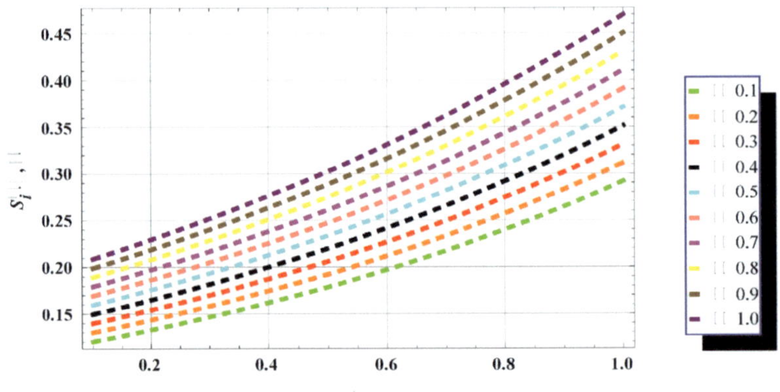

Figure 1.
Plot of Saturation $S^*_i(\xi,\theta)$ versus ξ for VIM Solution.

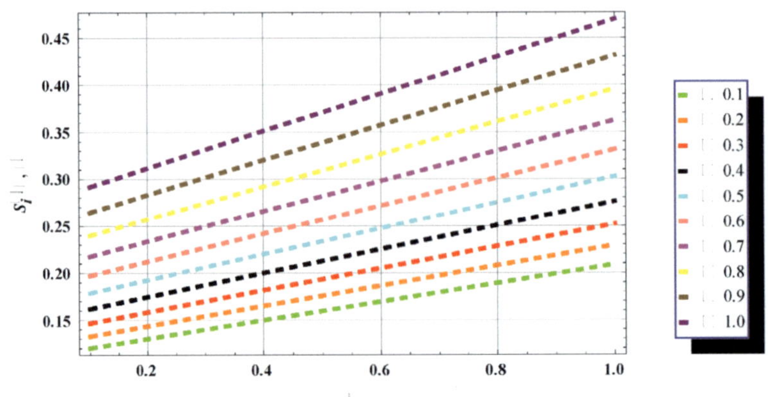

Figure 2.
Plot of Saturation $S^*_i(\xi,\theta)$ versus θ for VIM Solution.

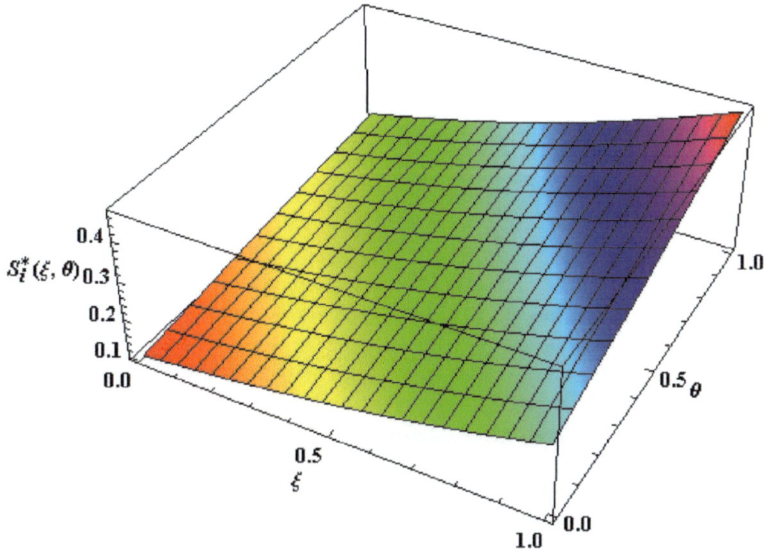

Figure 3.
3-Dimensional VIM Solution.

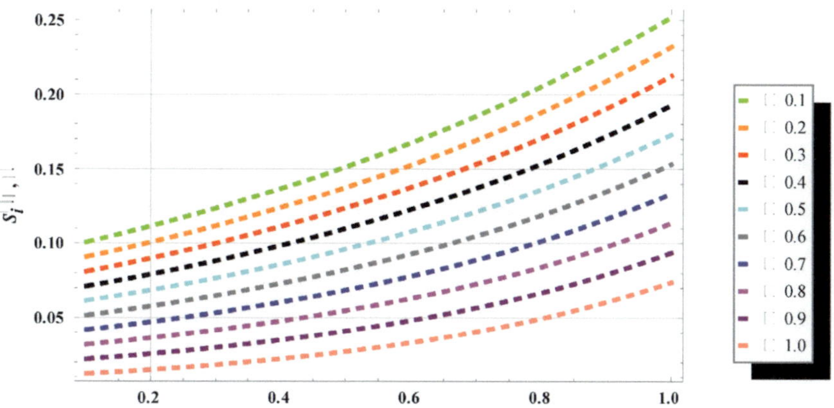

Figure 4.
Plot of Saturation $S_i^*(\xi, \theta)$ versus ξ for ADM Solution.

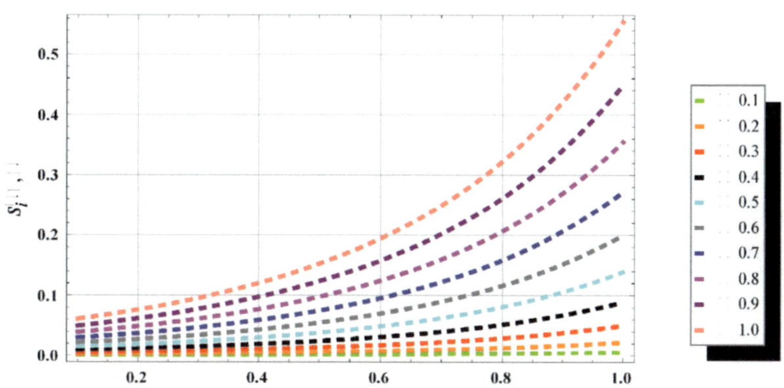

Figure 5.
Plot of Saturation $S_i^*(\xi, \theta)$ versus ξ for LADM Solution.

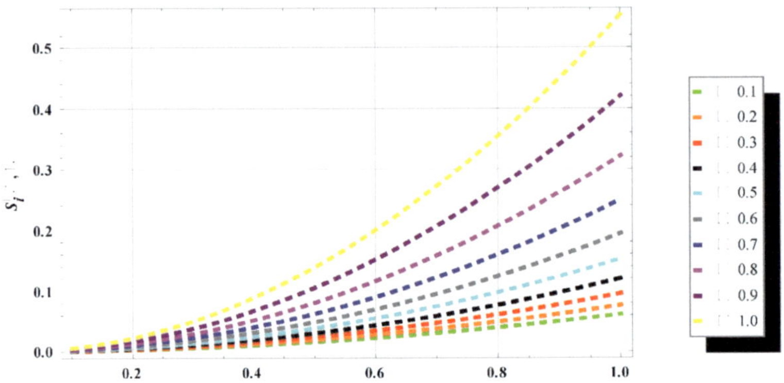

Figure 6.
Plot of Saturation $S_i^*(\xi, \theta)$ versus θ for LADM Solution.

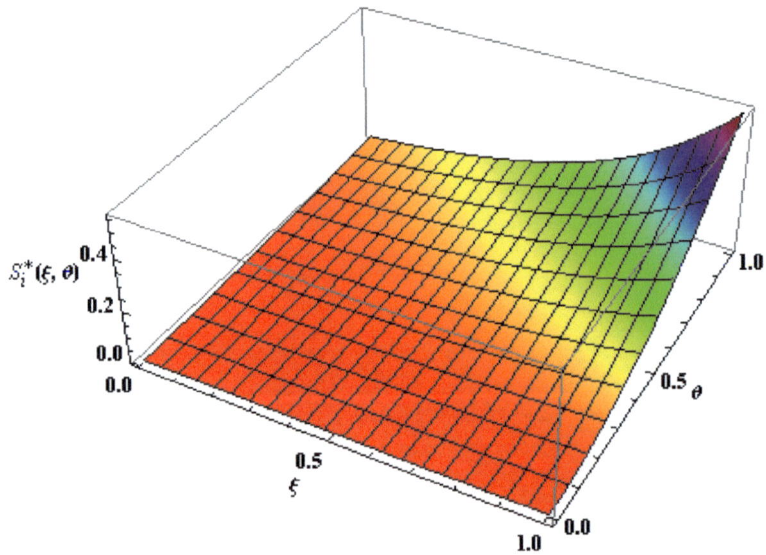

Figure 7.
3-Dimensional LADM Solution

saturation of injected water during imbibition, increases and it is noted that LADM gives faster accuracy compare to VIM and ADM (**Figures 1–7**).

5. Conclusions

The VIM, the ADM and the LADM are successfully applied to Burger's equation. The results which are obtained by ADM are a powerful mathematical tool to solve nonlinear partial differential equation. It has been noted that this method is reliable and requires fewer computations; and scheme LADM gives better and very faster accuracy in comparison with VIM.

Acknowledgements

The authors are thankful to Applied Mathematics and Humanities Department of S. V. National Institute of Technology, Surat for the encouragement and facilities.

Author details

Twinkle R. Singh
Applied Mathematics and Humanities Department, Sardar Vallabhbhai National Institute of Technology (SVNIT), Surat, Gujarat, India

*Address all correspondence to: twinklesingh.svnit@gmail.com

References

[1] He, J.H. (1998). Approximate solution of nonlinear differential equations with convolution product nonlinearities, Comput. Methods Appl. Mech. Engrg. 167, 69–73.

[2] Al-Hayani, W.(2011). Adomian decomposition method with Green's function for sixth- Orderboundary value problems, Comp. Math. Appl. 61, 1567–1575.

[3] Al-Hayani, W.and. Casasús, L (2005). Approximate analytical solution of fourth order Boundaryvalue problems, Numer. Algorithms 40, 67–78

[4] Abdou, M.A. and Soliman, A.A. (2005). New applications of variational iteration method, Physica D 211 (1-2), 1-8.

[5] Abdou, M.A. and Soliman, A.A. (2005). Variational iteration method for solving Burger's and coupled Burger's equations, J. Comput. Appl. Math. 181 (20), 245-251.

[6] Momani, S.,Abuasad, S. (2005). Application of He's varitional iteration method to Helmholtz equation, Chaos Solitons & Fractals 27, 1119-1123.

[7] Odibat, Z.M. , Momani,S.(2006). Application of variational iteration method to nonlinear differential equations of fractional order, Int. J. Nonlinear Sci. Numer. Simul. 7 (1), 27- 34.

[8] Soliman, A. A. (2006). Numerical simulation and explicit solutions of KdV–Burgers' and Lax's seventh-order KdV equations, Chaos Solitons & Fractals; 29 (2), 294-302.

[9] Scott, M. R. (1973). Invariant Imbedding and its Applications to Ordinary Differential Equations, Addison-vesley.

[10] Sweilam, N.Hand Khader, M.M. (2007). Variational iteration method for one dimensional nonlinear thermoelasticity, Chaos Solitons & Fractals, 32 (1),145-149.

[11] Adomian,G.91989). Nonlinear Stochastic Systems Theory and Applications to Physics, Kluwer Academic, Dordrecht.

[12] Adomian, G. (1994). Solving Frontier Problems of Physics: The Decomposition Method, Kluwer Academic, Dordrecht.

[13] Serrano, S.E. (2010). Hydrology for Engineers, Geologists, and Environmental rofessionals: An Integrated Treatment of Surface, Subsurface, and Contaminant Hydrology, Second Revised Edition, HydroScience, Ambler, PA.

[14] Adomian, G. (1983) R. Rach, Inversion of nonlinear stochastic operators, J. Math. Anal. Appl.91, 39–46.

[15] Adomian,G. (1983). Stochastic Systems, Academic, New York.

[16] Adomian, G.(1986). Nonlinear Stochastic Operator Equations, Academic, Orlando, FL.

[17] Adomian, G, Rach, R. (1993). A new algorithm for matching boundary conditions in decomposition solutions, Appl. Math. Comput. 58, 61–68.

[18] Bellman, R. E. and Adomian, G. (1985). Partial Differential Equations: New Methods for their Treatment and Solution, D.Reidel, Dordrecht.

[19] Bellomo, N.and Riganti, R. (1987). Nonlinear Stochastic System Analysis in Physics and Mechanics, World Scientific, Singapore and River Edge, NJ,

[20] Cherruault, Y. (1998). Modèles et méthodes mathématiques pour les

sciences du vivant,Presses Universitaires de France, Paris.

[21] Wazwaz, A. M. (1997). A First Course in Integral Equations, World Scientific, Singapore and River Edge, NJ.

[22] Wazwaz, A. M. (2002). Partial Differential Equations: Methods and Applications, A. A. Balkema,Lisse, The Netherlands.

[23] Wazwaz, A. M. (2009). Partial Differential Equations and Solitary Waves Theory, Higher Education Press, Beijing, and Springer-Verlag, Berlin.

[24] Wazwaz, A.M. (2011). Linear and Nonlinear Integral Equations: Methods and Applications,Higher EducationPress, Beijing, and Springer-Verlag, Berlin.

[25] Wazwaz, A.M. (2000). Approximate solutions to boundary value problems of higher order by themodified decomposition method, Comput. Math. Appl. 40, 679–691.

[26] Wazwaz, A.M. (2000). The modified Adomian decomposition method for solving linear and nonlinear boundary value problems of tenth-order and twelfth-order, Int. J. Nonlinear Sci. Numer. Simul. 1, 17–24.

[27] Wazwaz, A.M. (2000). A note on using Adomian decomposition method for solving boundary value problems, Found. Phys. Lett. 13, 493–498.

[28] Wazwaz, A.M. (2000). The numerical solution of special eighth-order boundary value problems by the modified decomposition method, Neural Parallel Sci. Comput. 8,133–146.

[29] Adomian, G.(1994). Solving Frontier Problems of Physics: Decomposition method.Kluwer, Boston, MA.

[30] Al-Sawalha, M.M., Noorani, M.S.M. and. Hashim, I. (2008). Numerical experiments on the hyperchaotic Chen system by the Adomian decomposition method, Int. J. Comput. Methods 5, 403–412.

[31] Bigi, D. and Riganti, R. (1986). Solution of nonlinear boundary value problems by the decomposition method, Appl. Math. Modelling 10, 49–52.

[32] Casasús L.and Al-Hayani, W. (2000). The method of Adomian for a nonlinear boundary value problem, Rev. Acad. Canar. Cienc. 12, 97–105.

[33] Duan, J.S. and Rach,R. (2011). New higher-order numerical one-step methods based on the Adomian and the modified decomposition methods, Appl. Math. Comput. 218, 2810–2828.

[34] Ebadi G. and Rashedi, S. (2010). The extended Adomian decomposition method for fourth order boundary value problems,Acta Univ. Apulensis 22, 65–78.

[35] He,J.H. (1997). A new approach to nonlinear partial differential equations, Comm. Nonlinear Sci. Numer. Simul. 2 (1997) 230–235.

[36] Hussain, M. and Khan, M. (2010). Modied Laplace decomposition method, Appl. Math. Sci, 36, 1769-1783.

[37] Jun-Sheng Duan a, Randolph Rach , Dumitru B˘aleanu , Abdul-Majid Wazwaz. (2012). A review of the Adomian decomposition method and its applications to fractional differential equations. Commun. Frac. Calc. 3 (2), 73 – 99.

[38] Jasem Fadaei. (2011). Application of Laplace – Adomian Decomposition Method on Linear and Nonlinear System of PDEs. Applied Mathematical Sciences, Vol. 5, 27, 1307 –1315.

[39] Jang, B. (2008). Two-point boundary value problems by the extended Adomian decomposition

method, J. Comput. Appl.Math. 219, 253–262.

[40] Khan, Y. (2009). An effective modification of the Laplace decomposition method for Nonlinear equations, International Journal of Nonlinear Sciences and Numerical Simulation, 10, 1373-1376.

[41] Lesnic, D. (2008). The decomposition method for nonlinear, second-order parabolic partial differential equations, Int. J. Comput. Math. Numeric. Simulat. 1, 207–233.

[42] Rach R.and. Baghdasarian, A . (1990). On approximate solution of a nonlinear differential equation, Appl. Math. Lett. 3, 101–102.

[43] Tatari, M.and Dehghan, M.(2006). The use of the Adomian decomposition method for solving multipoint boundary value problems, Physica Scripta 73, 672–676.

[44] Tsai, P.Y. and Chen, C.K. (2010). An approximate analytic solution of the nonlinear Riccati differential equation, J. Frank. Inst. 347, 1850–1862.

[45] Wazwaz A.M. and Rach, R. (2011). Comparison of the Adomian decomposition methodand the variational iteration method for solving the Lane-Emden equations of the first and second kinds, Kybernetes 40, 1305–1318.

[46] Wazwaz, A.M. (2012). A reliable study for extensions of the Bratu problem with boundary conditions, Math. Methods Appl. Sci. 35 (2012) 845–856.

[47] Khuri, S. A. (2001). A Laplace decomposition algorithm applied to a class ofNonlineardifferential equations," J. Math. Annl. Appl., 4, 141-155.

[48] Khuri, S. A. (2004). A new approach to Bratus problem," Appl. Math. Comp., 147 (2004) 31-136.

[49] Adomian, G.(1988). A review of the decomposition method in applied mathematics," Journal of mathematical analysis and applications, vol. 135, 501-544.

[50] Duan, J.S. and Rach, R. (2011). A new modification of the Adomian decomposition method for solving boundary value problems for higher order nonlinear differential equations, Appl.Math. Comput. 218, 4090–4118.

[51] Elgazery, S. Nasser. (2008). Numerical solution for the Falkner-Skan equation, Chaos, Solitons and Fractals, 35, 738 - 746.

[52] Fadaei, J. and Moghadam, M.M. (2012). Numerical Solution of Systems of Integral Differential equations by Using Modied Laplace Adomian Decomposition Method, World Applied Sciences Journal 19, 1818-1822.

[53] Ongun, M.Y. (2011). The Laplace Adomian Decomposition Method for solving a model for HIV infection of CD4+T cells, Math. Comp.Modell, 53, 597-603.

[54] Saei, F., Dastmalchi, F.,Misagh, D., Zahiri, Y.,Mahmoudi, M., Salehian, V. and Rafati Maleki, N. (2013). New Application of Laplace Decomposition Algorithm For Quadrtic Riccati Differential Equation by Using Adomian's Polynomials, Life Science Journal,10, 3s.

[55] Yusufoglu, E. (2006). Numerical solution of Dung equation by the Laplace decomposition algorithm, Appl. Math. Comput, 177, 572-580.

[56] Yusufoglu, Elcin,(Agadjanov), (2006). Numerical solution of Duffing equation by the Laplace decomposition algorithm, Appl. Math. Comput. 177, 572 - 580.

[57] Abbasbandy, S.(2006). Iterated He's homotopy Perturbation method for

quadratic Riccati differential equation, Applied Mathematics and Computation 175, 581-589

[58] Adomian G and Rach, R. (1993). Analytic solution of nonlinear boundary-value problems in several dimensions by decomposition, J. Math. Anal. Appl. 174,118–137.

[59] Al-Mazmumy, M.and Al-Malki,H. (2015). The modified Adomian Decomposition Method for solving Nonlinear Coupled Burger's Euations, Nonlinear Analysis and Differential Equations, Vol. 3, No.3, 111-122.

[60] Dehghan, M.and Tatari, M. (2010). Finding approximate solutions for a class of third-order non-linear boundary value problems via the decomposition method of Adomian, Int. J. Comput. Math. 87,1256–1263.

[61] Ebaid, A.E. (2010). Exact solutions for a class of nonlinear singular two-point boundary value problems: The decomposition method, Z. Naturforsch. 65a (2010) 1–6.

[62] Ebaid, A.E. (2011). A new analytical and numerical treatment for singular two-point boundary value problems via the Adomian decomposition method, J. Comput. Appl. Math. 235, 1914–1924.

[63] Geny, F.Z., Lin, I.Z., Cui, M.G. (2009). A piecewise variational iteration method for Riccati differential equations, Computers and mathematics with Applications 58, 2518-2522.

[64] Hashim, I., Adomian, G. (2006). Decomposition method for solving BVPs for fourth- order integro-differential equations, J. Comput. Appl. Math. 193, 658–664.

[65] He, J.H. (2004). Variational principle for some nonlinear partial differential equations with variable coefficients, Chaos Solitons Fractals 19, 847–851

[66] He, J.H. (2000). Variational iteration Method for autonomons ordinary differential systm, Applied Mathematics and Computation 114, 115-123.

[67] He, J.H.(1999). Variational iteration Method – a kind of non-linear analytical technique; some examples, International journal of Non-linear Mechanics 34, 699-708.

[68] He, J.H. (1998). Approximate analytical solution for seepage flow with fractional derivatives in porous media, Comput. Methods Appl. Mech. Engrg. 167, 57–68.

[69] Onur Kıymaz, Ayşegül Çetinkaya. (2010). Variational Iteration Method for a Class of Nonlinear Differential Equations. Int. J. Contemp. Math. Sciences, Vol. 5, 37:1819 – 1826

[70] Oroveanu,T.(1963). Scurgerea fluiidelor prin medii poroase neomogene, Editura Academiei Republicii Populare Romine,92, 328.

[71] Rach, R and Adomian, G. (1990). Multiple decompositions for computational convenience, Appl.Math. Lett. 3, 97–99.

[72] Sambath, M. and Balachandran, K. (2016). Laplace Adomian decomposition method for solving a fish farm model. Nonauton. Dyn. Syst. 3,104–111.

[73] Serrano, S.E. (2011). Engineering Uncertainty and Risk Analysis: A Balanced Approach to Probability, Statistics, Stochastic Modeling, and Stochastic Differential Equations, second Revised Edition, HydroScience, Ambler, PA.

[74] Taiwo, O. and Odetunde, O. (2010). On the numerical approximation of delay differential equations by a decomposition method," Asian Journal

of Mathematics and Statistics,vol. 3, pp. 237-243.

[75] Verma, A.P. (1970): Perturbation solution in imbibition in a cracked porous medium of small inclination, IASH, 13, 1, 45-51.

[76] Elgazery, S.N. (2008). Numerical solution for the Falkner-Skan equation, Chaos, Solitons and Fractals, 35,738-746.

[77] Graham, R.A. and Richardson, J.G. (1959): Theory and applications of imbibition phenomena in recovery of oil, Trans, Aime, 216.

[78] Scheidegger, A.E. (1960). The Physics of Flow through Porous Media, University of Toronto press.

[79] Cole, J. D. (1951): On a quasilinear parabolic equation occurring in aerodynamics. Q. Appl. Math. g,225-236.

[80] Hopf, E. (1950): The partial differential equation ut + uux = ε uxx comm. Pure Appl. Math. 3, pp – 201-230.

[81] Tamer A. Abassy, Magdy A. El-Tawil, H. Fl. Zoheiry, Towards a modified Variational iteration method, Journal of Computational and Applied mathematics 207 (2007) 137-147.

[82] Tamer A. Abassy, Improved Adomian decomposition method; Computers and Mathematics with Applications 59 (2010) 42-54.

Section 3

Experimental Challenges in Geological Applications of Porous Media

Chapter 4

Dissolution Mass Transfer of Trapped Phase in Porous Media

Anindityo Patmonoaji, Yingxue Hu, Chunwei Zhang and Tetsuya Suekane

Abstract

Dissolution mass transfer of trapped phase (TP) to flowing phase (FP) in porous media plays significant roles in hydrogeology, e.g., groundwater contamination by non-aqueous phase liquids, groundwater in-situ bioremediation, and geological carbon sequestration. In this chapter, this phenomenon is described. First, the physical and mathematical models are given. Afterwards, various conditions affecting this process, i.e., porous media characteristics, capillary trapping characteristics, flow bypassing, TP characteristics, and FP velocity, are discussed. These various conditions are described based on three parameters affecting the dissolution mass transfer: TP interfacial area ($\boldsymbol{A}$), TP dissolution ratio ($\boldsymbol{\xi}$), and mass transfer coefficient ($\boldsymbol{k}$). Eventually, models to predict the mass transfer are formulated based on non-dimensional model. All of the data in this chapter are based on the experiments obtained by using micro-tomography and a series of image processing techniques from our latest works.

Keywords: dissolution, mass transfer, porous media, micro-tomography, interfacial area, trapped phase cluster

1. Introduction

1.1 Applications of dissolution mass transfer in porous media

Dissolution mass transfer of trapped phase (TP) to flowing phase (FP) in porous media plays significant roles in hydrogeology (**Figure 1**), e.g., groundwater contamination by non-aqueous phase liquid (NAPL) [1–3], groundwater in-situ bioremediation [4, 5], and geological carbon sequestration (GCS) [6].

In groundwater contamination by NAPL, NAPL could leak from industrial plant and contaminate the groundwater stream [1–3]. When the NAPL enters the groundwater, it displaces the existing groundwater, and then it was trapped due to capillary trapping. As it is trapped, mass transfer from the trapped NAPL to the groundwater flow occurs, transferring the dissolved NAPL to the groundwater. As a result, groundwater is contaminated, and the period of contamination depends on the mass transfer rate of trapped NAPL to the groundwater. Low mass transfer rate results in a prolong contamination, and vice versa.

In-situ bioremediation is one of the decontamination method of NAPL contamination [4, 5]. This method harnesses the ability of native bacteria in the groundwater to biodegrade the contaminants. To enhance the biodegradation process, O_2,

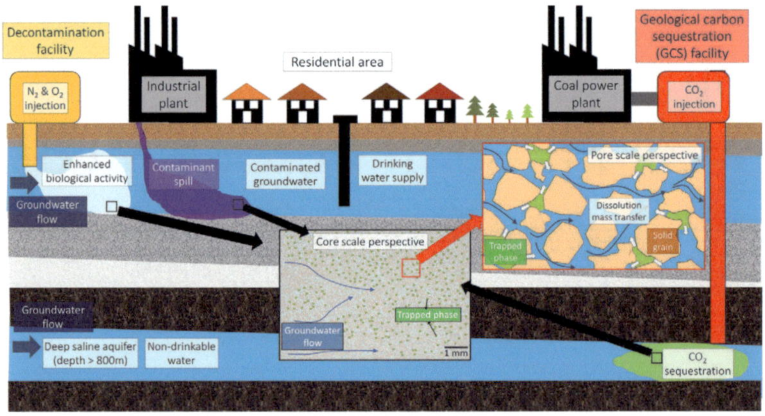

Figure 1.
Dissolution mass transfer in porous media phenomena in hydrogeology (modified from Patmonoaji et al. [7] with permission).

CH_4, NH_3, CO_3 and N_2 gases are injected at the upstream of the contamination area to supply nutrients and stimulate the bacteria. After the injection, these gases are trapped due to capillary trapping and thus mass transfer occurs. The mass transfer rate is important because it affects the efficiency of the gas transport. Higher mass transfer rate means faster delivery but faster depletion of trapped gases. Therefore, it should be controlled to match the bacteria needs.

In GCS, the captured CO_2 from the point source, e.g., coal power plant, steel factory, and cement factory, is injected into deep saline aquifers for sequestration. After the injection, the CO_2 is trapped in the groundwater due to capillary trapping [6, 8]. Although the CO_2 can be sequestered this way, the reservoir pressure increases due to the injection. However, by the time, the trapped CO_2 will dissolve, reducing the reservoir pressure. As a result, the mass transfer rate is critical to approximate the pressure reduction rate for safer CO_2 sequestration process.

Given the scale of the phenomena is massive, the observation scale can be categorized into pore-scale, core-scale, and field-scale (**Figure 1**). Pore-scale is in the order of μm, and core-scale is in the order of cm. Field-scale, however, is in the order of m to km. Each of the scale is used to explain the scales above it. Pore-scale and core-scale are performed in a laboratory, whereas field-scale is performed in the field. Only pore-scale and core-scale will be discussed in this chapter.

Given the opaqueness of porous media system, direct observation of dissolution process is difficult to be performed. To evaluate the mass transfer process in detail, the interfacial area ($\boldsymbol{A}$) of TP needs to be measured. With the information of interfacial area of TP, the mass transfer coefficient ($\boldsymbol{k}$) can be calculated (described later in Eq. (5)). In addition, the distribution of the TP clusters is also need to be monitored to assess the dissolution mass transfer process.

Earlier studies mainly relied only on core-scale studies by the measurement of dissolved phase concentration from the effluent and amount of trapped phase through gravimetrical measurement without knowing the trapped phase characteristics inside [1–5]. However, because the trapped phase characteristics are required to measure the $\boldsymbol{A}$ for the mass transfer, and its characteristics are strongly controlled by the porous media characteristics, the studies become phenomenological because $\boldsymbol{k}$ cannot be calculated. In addition, the overall dissolution process cannot be observed, and thus flow bypassing cannot be monitored although it also affects the mass transfer process.

In this chapter, dissolution mass transfer in porous media is discussed based on experimental observation on the dissolution mass transfer of various trapped gas into flowing water in porous media by using micro-tomography and image processing [7, 9–12]. By using this methods, $\boldsymbol{A}$ and distribution of TP clusters can be monitored throughout dissolution process, and thus $\boldsymbol{k}$ can be calculated. In addition, by monitoring dissolution process, flow bypassing was investigated, and thus its effect to $\boldsymbol{k}$ was found. Moreover, unique dissolution process of TP with high dissolution ratio ($\boldsymbol{\xi}$), which is the ratio of solubility limit ($\boldsymbol{C_{sol,i}}$) and density ($\boldsymbol{\rho_i}$) was observed. Its effect to $\boldsymbol{k}$ was also found. Therefore, all of three parameters effecting dissolution mass transfer ($\boldsymbol{A}$, $\boldsymbol{\xi}$, and $\boldsymbol{k}$), were elucidated.

The contents of this chapter are described as follow. In Section 2, the physical and mathematical model are given. In Section 3, the general description about the micro-tomography and image processing methods are described. In Section 4, various parameters affecting this process, i.e., porous media characteristics, capillary trapping characteristics, flow bypassing, TP properties, and FP velocity, are discussed. All of these condition will be investigated based on the point of view of three parameters affecting dissolution mass transfer: $\boldsymbol{A}$, $\boldsymbol{\xi}$, and $\boldsymbol{k}$. Porous media together with capillary trapping characteristics affect the amount of $\boldsymbol{A}$ for mass transfer process. Flow bypassing affects the uniformity of mass transfer process and eventually $\boldsymbol{k}$. TP properties affect the $\boldsymbol{\xi}$ through its $\boldsymbol{C_{sol,i}}$ and $\boldsymbol{\rho_i}$. Lastly, FP velocity affects the advection rate of the solute, influencing $\boldsymbol{k}$. Eventually, the mass transfer model based on non-dimensional number is also formulated at the end of the chapter.

2. Physical and mathematical model

2.1 What is dissolution mass transfer?

At first, we can start from the basic by describing mass transfer. Mass transfer is a process of a movement of mass from one medium to another medium. The examples can be found easily in the surrounding of our daily lives. For example, the mass transfer of camphor in a bathroom into the surrounding air to improve the smells of the room (**Figure 2a**). Another example is the mass transfer of oxygen from the air into a lake (**Figure 2b**). Without the supply of oxygen through mass transfer, the fishes living in the lake cannot breathe the oxygen and stay alive.

Similar with other transport phenomena process, i.e., heat transfer and momentum transfer, mass transfer occurs because of non-equilibrium condition [13–15]. In case of mass transfer, equilibrium can be achieved once the system possesses equal chemical potential ($\boldsymbol{\eta}$) of the considered component ($\boldsymbol{i}$):

$$\eta_{i,1} = \eta_{i,2} \tag{1}$$

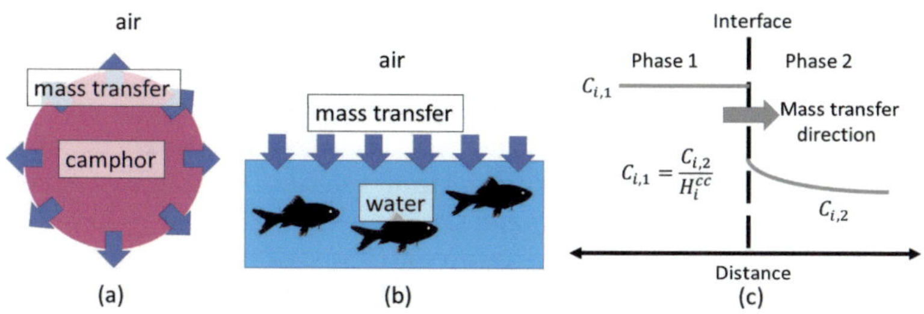

Figure 2.
(a) Mass transfer of camphor to surrounding air, (b) mass transfer of oxygen from air to water in lake, and (c) illustration of mass transfer process.

with subscripts 1 and 2 represent the phases. This equation can also be described with partition coefficient or Henry's coefficient (**Figure 2c**) as follow:

$$C_{i,1} = \frac{C_{i,2}}{H_i^{cc}} \tag{2}$$

with C_i as concentration and H_i^{cc} as dimensionless Henry coefficient of component i, which is constant with pressure but varies with temperature.

The concept of mass transfer nowadays is mainly based on the works of Thomas Graham (1805–1869) and Adolf Fick (1829–1901). Thomas Graham was the pioneer in the study of mass transfer through his works in the diffusion of H_2 gas in a tube [16]. By using analogies with Fourier's work [17] in heat conduction, Adolf Fick formulated the based equation for mass transfer that is the Fick's law of diffusion mass transfer [18] defined as follow:

$$j = -D_i \nabla C_i \tag{3}$$

with j as mass flux and D_i as diffusion mass transfer coefficient of compound i in the medium. Another form is based on Newton's law of cooling as follow:

$$j = k\Delta C_i \tag{4}$$

with k as mass transfer coefficient. In this approach, k and ΔC_i is used instead of D_i and ∇C_i. As a result, the mass transfer is judged based on concentration difference without knowing the concentration gradient. Because the concentration gradient in this system is difficult to obtain in the system, this form of mass transfer equation is more preferable in mass transfer from an interface into bulk solution, such as in the hydrogeology processes mentioned in Section 1.

In case of dissolution mass transfer, an additional process present. Dissolution here refers to the decrease in volume due to mass transfer. As a result, the Eq. (4) can be modified to volume perspective as follow:

$$\rho_i \frac{dV}{dt} = kA\,(C_{sol,i} - C_i) \tag{5}$$

with ρ_i as density of compound i, V as volume, and t as time. Therefore, high mass transfer rate also results in high volume decrease rate or dissolution rate. The example of camphor in **Figure 2a** provides an example of dissolution mass transfer, whereas oxygen mass transfer in a pool of water in **Figure 2b** provides an example of mass transfer without dissolution.

2.2 Physical model of mass transfer from an interface into a bulk solution

There are several mechanisms models explaining mass transfer from an interface into a bulk solution [14]. Due to the analogy between mass transfer and heat transfer, similar models with heat transfer are also used. However, because most of heat transfer application is in fluid–solid system, whereas mass transfer applications occur in fluid–fluid system, not all model can be used for both.

The simplest physical model is the stagnant film theory (**Figure 3a**) [19] that assumes the existence of a stagnant film near the interface of mass transfer process. This stagnant film is also referred as an unstirred layer because no fluid motion exists in this film. However, this film is hypothetical because fluid motions still occur beyond the no-slip condition at interface. The effects of velocity and other

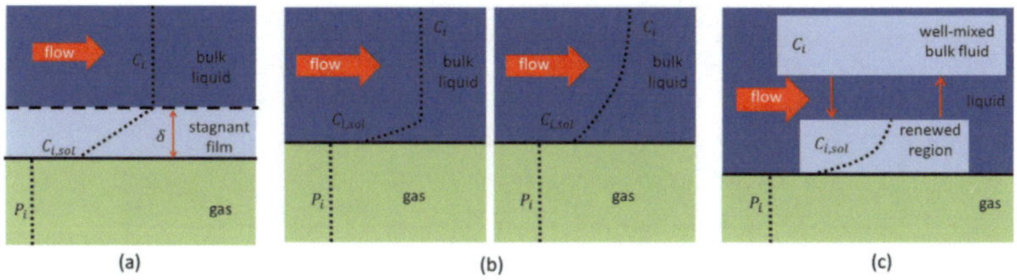

Figure 3.
(a) Illustration of stagnant film model, (b) penetration theory model at low and high velocity, and (c) surface renewal theory model.

parameters are then represented by the thickness of this stagnant film. Higher mass transfer rate results in thinner δ, and vice versa.

Second mechanisms model is penetration theory (**Figure 3b**) [20]. The key assumption is that a thick moving film is continuously generated in the bulk fluid, which the mass transfer occurs by diffusion across this film. The effect of velocity to the mass transfer is represented by the concentration distribution. As the solute diffuse into the bulk liquid, it will be flushed away by the flow. Higher velocity results in faster removal of the solute. As a result, higher velocity results in steeper concentration distribution, leading to higher mass transfer rate, and vice versa.

Third mechanisms model is surface renewal theory (**Figure 3c**) [21], which is the further development of penetration theory. The bulk liquid is depicted into two regions, a region near the interface that is renewed quickly and a region of well-mixed bulk fluid. The renewed region is renewed by the flow and similar with penetration theory. However, it is constantly exchanged with new elements from a second bulk region, making this theory closer to the realistic physical situation.

Among these models, however, no theory is completely satisfactory. Nevertheless, these models provide valuable insight to the mechanisms in mass transfer. For more detail, readers are advised to check the articles in the references [19–21].

For more empirical modeling of the mass transfer process, an empirical equation based on dimensionless number is usually used. The experimental data of mass transfer is represented by power-law relation of dimensionless number as follow:

$$\boldsymbol{Sh} = \epsilon_1 \boldsymbol{Re}^{\epsilon_2} \tag{6}$$

with ϵ_1, ϵ_2, and ϵ_3 as constants, $\boldsymbol{Sh}$ as Sherwood number representing the mass transfer rate and $\boldsymbol{Re}$ as Reynolds number representing the bulk fluid velocity and geometry of the system. These non-dimensional number are defined as follow:

$$\boldsymbol{Sh} = \frac{\boldsymbol{kL}}{\boldsymbol{D_i}} \tag{7}$$

$$\boldsymbol{Re} = \frac{\rho_{FP} \boldsymbol{UL}}{\mu_{FP}} \tag{8}$$

with ρ_{FP} as FP density, μ_{FP} as FP absolute viscosity, $\boldsymbol{U}$ as FP velocity, and $\boldsymbol{L}$ as characteristic length.

In case of porous media, the physical mechanism of mass transfer is actually similar. In an open space, when dissolution occurs, the solute can be flushed away easily. In porous media, on the other hand, these processes are governed by the solute transport in porous media. In addition, in the application of hydrogeology, the TP clusters are distributed throughout the porous media (**Figure 4**), resulting in continuous mass transfer process along the porous media. As a result, a special attention to the solute transport throughout the porous media is required before assessing the mass transfer process with Eq. (5).

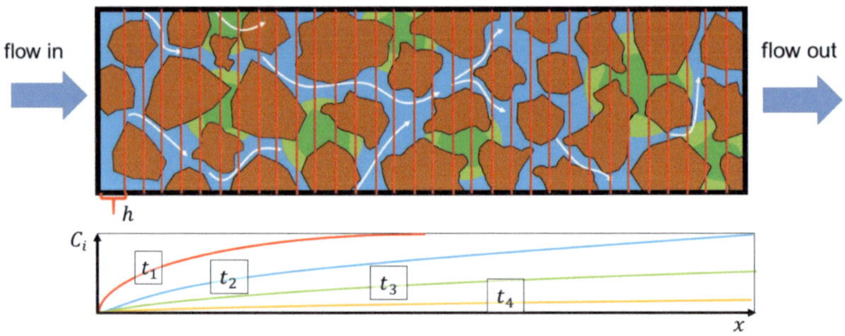

Figure 4.
Illustration of dissolution mass transfer in porous media and the predicted $\mathbf{C}_{i,x,t}$ *based on equation 16.*

2.3 Fluid flow and solute transport in porous media

The fluid flow in hydrogeology application is relatively slow and can be classified as Stokes flow or creeping flow, which the ratio between fluid velocity and viscosities is relatively low. This condition is represented by $\boldsymbol{Re}$ lower than one. Steady-state Stokes equation is given as follow:

$$\mu_{FP}\nabla^2 \boldsymbol{U} = \nabla \boldsymbol{P} - \rho_{FP}\boldsymbol{g} \tag{9}$$

with $\boldsymbol{P}$ as pressure and $\boldsymbol{g}$ as gravitational acceleration. This equation is then further derived into Darcy law, which is introduced by Henry Darcy [22] as a relationship to describe fluid flux ($\boldsymbol{q}$) in sand filters, which was first proposed by Nutting [23] and Wyckoff et al. [24], described as follow:

$$\boldsymbol{q} = \frac{\boldsymbol{K}}{\mu_{FP}}(\nabla \boldsymbol{P} - \rho_{FP}\boldsymbol{g}) \tag{10}$$

with $\boldsymbol{K}$ as permeability. Given its original development, Darcy equation has been used mainly in the field of fluid flow in porous media. Similarly, with Stokes flow, to satisfy Darcy regime, the $\boldsymbol{Re}$ needs to be low. Experimental tests demonstrated that it can still be achieved when the $\boldsymbol{Re}$ is lower than ten [25].

For the solute transport in porous media, it is affected by the tortuosity and velocity profile in the pore networks. Nevertheless, under Darcy regime, it can be predicted by using advection-dispersion equation [26] as follow:

$$\frac{\partial \boldsymbol{C}_i}{\partial \boldsymbol{t}} = \boldsymbol{D}_{hdl}\nabla^2 \boldsymbol{C}_i - \mathrm{q}\nabla \boldsymbol{C}_i - \boldsymbol{R} \tag{11}$$

with, $\boldsymbol{D}_{hdl}$ as hydrodynamic dispersion coefficient and $\boldsymbol{R}$ as external mass transfer source or sink. With this equation, solute transport along the porous media can be predicted, and it can be used to assess the mass transfer with Eq. (5).

2.4 Calculation of dissolution mass transfer in porous media

To evaluate the dissolution mass transfer in porous media, combination of modified version of Eqs. (5) and (11) are used [1–5, 9]. To calculate the mass transfer coefficient in porous media, Eq. (5) is slightly modified as follow:

$$\rho_i\phi\frac{d\boldsymbol{S}_{TP}}{d\boldsymbol{t}} = \boldsymbol{ka}_{TP}\left(\boldsymbol{C}_{sol,i} - \boldsymbol{C}_i\right) \tag{12}$$

with ϕ as porosity, S_{TP} as TP saturation, and a_{TP} as specific interfacial area, which is the amount of A per porous media volume. However, the C_i, which is the driving force of the mass transfer, is unknown. To approximate it, C_i can be calculated using Eq. (11). Because the experiment can be assumed as a one-dimensional column, one-dimensional advection-dispersion equation as follow:

$$\phi(1 - S_{TP,cs})\frac{\partial C_i}{\partial t} = \frac{\partial}{\partial x}\left[\phi(1 - S_{TP,cs})D_{hdl}\frac{\partial C_i}{\partial x}\right] - q\frac{\partial C_i}{\partial x} - \rho_i\phi\frac{\partial S_{TP,cs}}{\partial t} \tag{13}$$

with $S_{TP,cs}$ as cross-sectional average of TP saturation and x as axial direction, can be used. In addition, it can be simplified with some assumptions [1–3, 9]. First, because dissolution mass transfer is a prolong process, the problem can be approached as a pseudo-steady state, neglecting the time accumulation term at left-hand side of the equation. Second, the dispersion term, which is the first term at the right-hand side of the equation, can be neglected due to the dominance in advection. It can be assessed by calculating the Peclet number (Pe) as follow:

$$Pe = \frac{Ud_{par,50\%}}{D_i} \tag{14}$$

with $d_{par,50\%}$ as median particle size. Pe represents the ratio of advection and diffusion of solute transport. High Pe number represents advection dominant regime, whereas low Pe represents diffusion dominant regime. When the value of Pe is higher than five, it represent advection dominant regime [27]. Thus, the dispersion term can be neglected. This was also proven with order of magnitude analysis [3, 9]. The order of magnitudes of accumulation and dispersion terms are significantly lower than the others. Hence, it can be simplified to:

$$q\frac{\partial C_i}{\partial x} = -\rho_i\phi\frac{\partial S_{TP,cs}}{\partial t} \tag{15}$$

By discretizing ∂x to Δx, C_i at the location of $x + \Delta x$ and at the given time of t ($C_{i,x+\Delta x,t}$) as shown in **Figure 4**, can be calculated as follow:

$$C_{i,x+\Delta x,t} = C_{i,x,t} - \frac{\rho_i\phi}{q}\frac{\Delta S_{TP,cs}}{\Delta t}\Delta x \tag{16}$$

The average of concentration in between x and $x + \Delta x$ at the time of t ($C_{i,x_{avg},t}$) (**Figure 4**) then can be calculated by averaging the $C_{i,x,t}$ and $C_{i,x+\Delta x,t}$ as follow:

$$C_{i,x_{avg},t} = \frac{C_{i,x,t} + C_{i,x+\Delta x,t}}{2} \tag{17}$$

This $C_{i,x_{avg},t}$ then can be used to calculate the mass transfer coefficient as follow:

$$k = \frac{-\rho_i\phi\frac{\Delta S_{TP,cs}}{\Delta t}}{a_{TP,cs,t_{avg}}\left(C_{i,sol} - C_{i,x_{avg},t}\right)} \tag{18}$$

with $a_{TP,cs,t_{avg}}$ as the cross-sectional average of a_{TP} that is averaged in between the time of $t - \Delta t$ ($a_{TP,cs,t-\Delta t}$) and t ($a_{TP,cs,t}$) calculated as follow:

$$a_{TP,cs,t_{avg}} = \frac{a_{TP,cs,t} + a_{TP,cs,t-\Delta t}}{2} \tag{19}$$

For the modeling of mass transfer, $\boldsymbol{Re}$ and $\boldsymbol{Sh}$ in Eqs. (7) and (8) were used with $\boldsymbol{L}$ equal to $\boldsymbol{d_{par,50\%}}$.

3. Experimental procedure

3.1 Experimental setup

As mentioned in Section 1, this chapter is based on the experimental observation by using micro-tomography and image processing [7, 9–12]. For the micro-tomography, an X-ray microtomography scanner (micro-CT) was used. This micro-CT is able to generate 3D images in the form of 992 cross section slice image with the dimension of 992 × 992 voxels with the voxel size of 16.427 μm. As a result, observation area with the size of 16.34 × 16.34 × 16.34 mm^3 was generated.

To generate the porous media, cylindrical container with an inner diameter of 10 mm was used as the casing of porous media. Inside this container, various particles will be packed inside to generate various porous media. All of the packing was performed under vibration to generate closed random packing classification [28]. As a result, the quality of the packing can be controlled. Additionally, sintered glass discs were placed at inlet and outlet to generate uniform flow.

To generate capillary trapping and similar condition with hydrogeology conditions in Section 1, the porous media will be sequentially saturated with degassed deionized (DDI) water and the gas. First, the gas was injected into the porous medium to saturate the porous media and remove all of other gases. Second, DDI water was injected upward to saturate the porous media, generating a condition similar with groundwater. Third, the experimental gas was injected again to remove most of the DDI water, resulting in similar condition with gas injection into groundwater. Finally, the DDI water was injected upward at the desired experimental flow rate. With this method, the generated trapped gas is uniform along the sample. Once the DDI water had breakthrough, and no gas flow was observed in the outlet tube, creating a quasi-steady state, the experiment began, and the dissolution process was monitored with micro-CT. The schematic about the experimental setup is given in **Figure 5**.

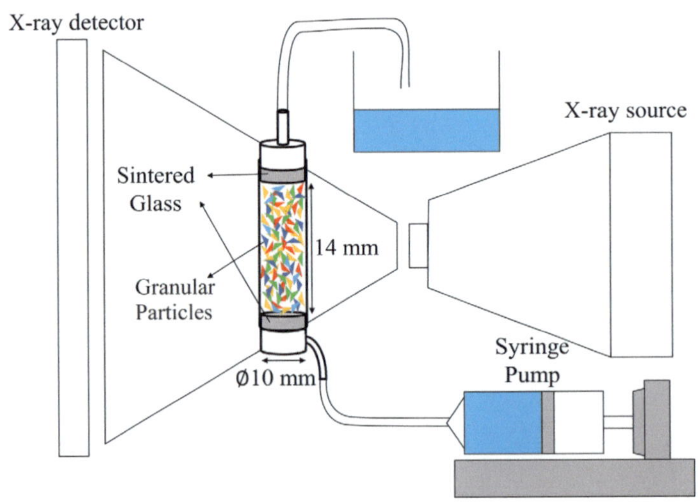

Figure 5.
Schematics of experimental setup with micro-tomography (modified from Patmonoaji et al. [9] with permission).

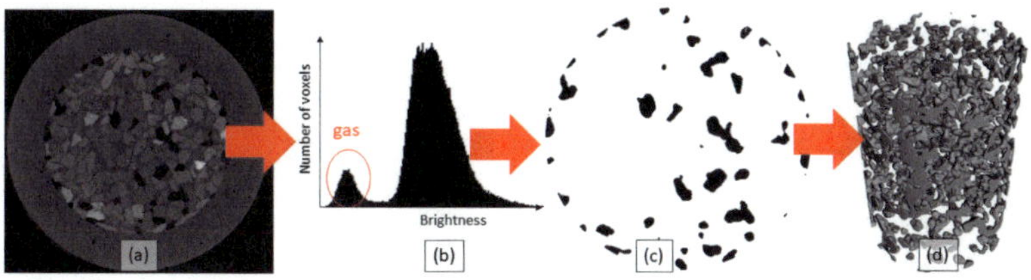

Figure 6.
Binarization image processing methods: (a) cross sectional image of the porous media, (b) histogram of the gray value, (c) binarized image with black as the gas, and (d) 3D binarized image of gas (modified from Patmonoaji et al. [1] with permission).

3.2 Image processing

For the data processing, image processing techniques were carried out to quantify the data from the micro-CT image. The main method is to differentiate the gas from the surroundings by performing image binarization (**Figure 6**). Although the contrast between water and particles is indistinguishable, the contrast of the gas among other phases is easily distinguishable. Therefore, a direct thresholding method was enough to separate gas from other phases. With this binarized images, then quantitative data, i.e., amount of trapped gas, and surface area of the trapped gas, can be simply derived with image processing to assess the dissolution mass transfer. Furthermore, by identifying separated trapped gas clusters, each of the trapped gas clusters can be characterized based on its volume and surface area. In addition, by performing similar method to porous media saturated with gas, porosity of the porous media can be obtained. Moreover, by applying watershed (WS) segmentation method [11], the throats and pores of the porous media can be identified and measured.

4. Factors affecting dissolution mass transfer in porous media

4.1 Porous media characteristics

A porous medium can be described as a solid containing void networks inside it. Structure-wise, the void network can be seen as a series of pores and throats (**Figure 7a**) [28, 29]. Pores are connected by a throat, which is a constriction in the network. Because capillary trapping mainly occupies pores, pore characteristics affect the morphology of the capillary trapping that eventually controls the amount A for the mass transfer process (**Figure 7b**).

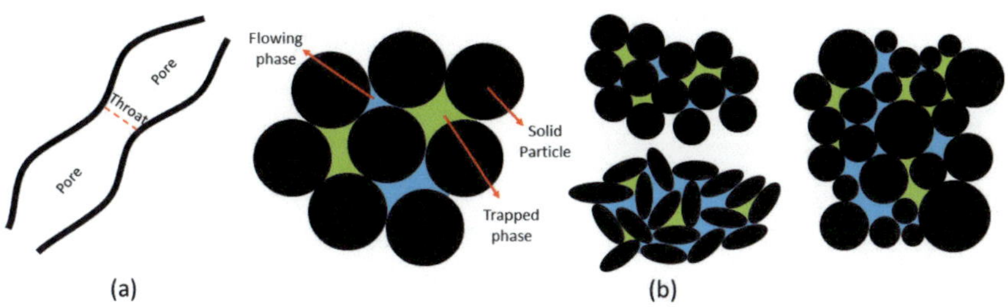

Figure 7.
(a) Illustration of pore and throat in pore network and (b) effect of porous media characteristics to TP morphology.

These pore and throat in porous media are mainly governed by the shape, size, and distribution of the granular particles forming the porous media. Shown in **Table 1** are the properties of nine granular particles used to generate porous media [10, 11]. Those properties C_u as particle distribution uniformity coefficient, C_c as particle distribution gradation coefficient, and Φ as porosity. Three particle shapes, which are plastic beads (PB), glass beads (GB), and Toyoura sands (TS), are used. PB represents angular shape particle like quartz sands, whereas GB represents sphere particles. TS, on the other hand, represents natural silica sands.

Figure 8a shows the pore size distribution (PSD) of the porous media generated from those particles. The PSD is shown as probability density function (PDF) of pore equivalent diameter (d_{pore}). To obtain this PSD, combinations of micro-tomography and WS method were performed [11]. GB tends to generate larger d_{pore} than PB, whereas TS generate similar d_{pore} with GB. For particle size, larger particles generate larger d_{pore}. However, in case of particle size distribution as compared by PB 180–212 μm, PB 125–300 μm, and PB 125–710 μm, similar PSD was found. Although the range of particle size distribution is different, the $d_{par,50\%}$ shown in **Table 1**, is actually similar. Pores are voids generated due to contact of granular particles (**Figure 7b**). Contact of large particles generates large voids. However, if the distribution of particles is wide. Large voids that is generated by large particles can be filled with the smaller particles. As a result, d_{pore} tends to be controlled by the smaller particle size, particularly $d_{par,50\%}$. This is also consistent with the Φ.

Type	Size range (μm)	$d_{par,50\%}$ (μm)	C_u	C_c	Φ
PB	125–300	184	1.5	0.9	0.43
	125–710	212	1.8	0.8	0.41
	180–212	196	1.1	1.0	0.44
	355–425	390	1.1	1.0	0.45
	600–710	655	1.1	1.0	0.44
GB	180–212	196	1.1	1.0	0.33
	355–425	390	1.1	1.0	0.34
	600–710	655	1.1	1.0	0.35
TS	180–212	196	1.1	1.0	0.32

Table 1.
Properties of nine porous media generated from different shape, size, and distribution of granular particles.

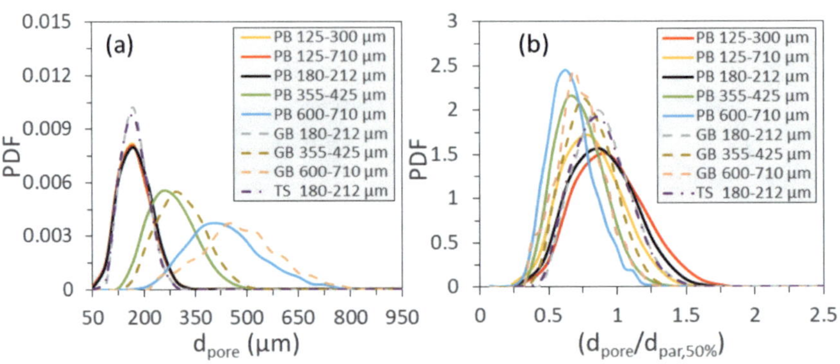

Figure 8.
(a) PSD and (b) $d_{par,50\%}$-normalized PSD.

Wider particle range generates lower $\boldsymbol{\Phi}$ because the large pores generated from the contact of large particles are filled by smaller particles.

Proportionality of the PSD is also found by normalizing PSD with the $d_{par,50\%}$ (**Figure 8b**). By normalizing the PSD with $d_{par,50\%}$, the PSD tend to converge into a single curve but not exactly. The differences are due to particle shape variation. Nevertheless, it demonstrates proportionality to $d_{par,50\%}$ for each particle shape.

The pore shape can also be evaluated through its surface area (A_{pore}) and volume (V_{pore}) as shown in **Figure 9**. Each point is the result obtained from averaging A_{pore} of all pores in a log-scale range of V_{pore}. Typical pores shape detected in porous media with PB, GB, and TS is also given as yellow, blue, and green objects, respectively. As shown in normal-scale, pores in smaller $d_{pore,50\%}$ generates larger A_{pore}. Related with particle shape, the A_{pore} of PB tends to be higher than GB and TS. GB generates pore with the smallest A_{pore}, whereas TS generate pore with A_{pore} in between PB and GB. This higher A_{pore} is due to the angular shapes of PB.

4.2 Capillary trapping characteristics

Capillary, buoyancy, and viscous forces play important roles in the formation of capillary trapping (**Figure 10a**). These forces can be represented with capillary number (Ca) and Bond number (Bo) as follow:

$$Ca = \frac{U\mu_{WP}}{\gamma} \tag{20}$$

$$Bo = \frac{\Delta\rho g {d_{par,50\%}}^2}{\gamma} \tag{21}$$

Ca represents the comparison between viscous and capillary forces, whereas Bo represents the comparison between buoyancy and capillary forces. Buoyancy and viscous forces exert on the TP to escape from the trapped location, whereas capillary force resist these forces to keep the TP in the pore. If Bo and Ca is higher than 0.35 and 10^{-3}, respectively, no capillary trapping occurs due to excessive buoyancy and viscous forces [30, 31]. In hydrogeology applications, the Bo and Ca are much lower, resulting in the possibility of capillary trapping generation.

Wettability of the porous media also affects the TP clusters morphology (**Figure 10b** and **c**). In case the TP is non-wetting phase (NWP), the TP is located in

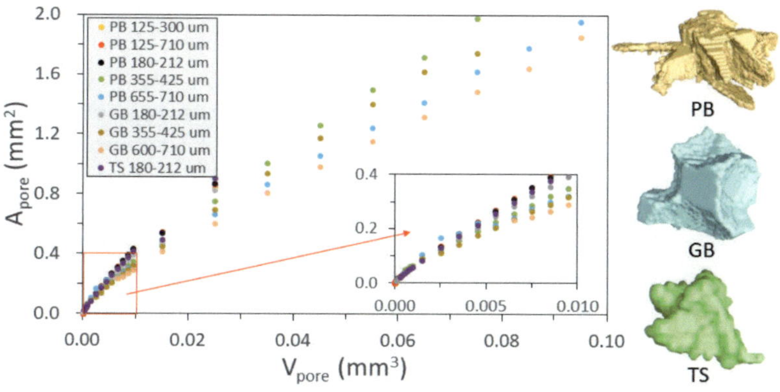

Figure 9.
V_{pore} and A_{pore} in the porous media and the typical pore shapes in each particle shape (modified from Patmonoaji et al. [11]).

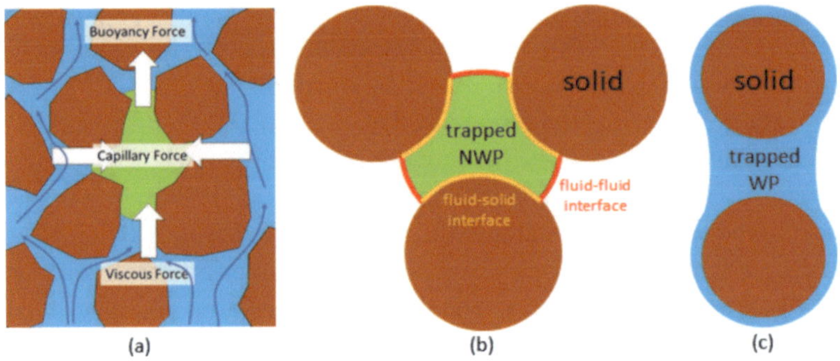

Figure 10.
(a) Illustration of forces exerting on TP in porous media and morphology of (b) trapped NWP and (c) WP in porous media.

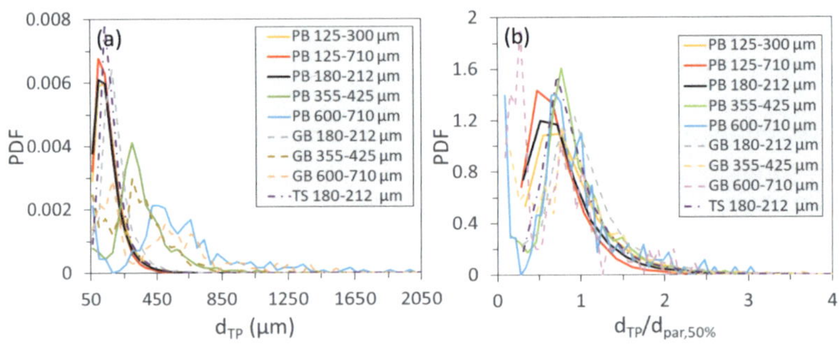

Figure 11.
(a) TPSD and (b) $d_{par,50\%}$*-normalized TPSD (modified from Patmonoaji et al. [10]).*

the center of the pore. On the other hand, in case the TP is wetting phase (WP), the TP tends to surround the solid. In this work, we mainly focus on the former condition.

The trapped phase size distribution (TPSD) in the porous media described in **Table 1** is given in **Figure 11** as PDF of TP equivalent diameter (d_{TP}). As shown in **Figure 11a**, with the increase in $d_{par,50\%}$, d_{TP} increases. Compared with the PSD, the TPSD tends to be larger due to the possibility to form multi-pores bubbles. By normalizing with $d_{par,50\%}$ as shown in **Figure 11b**, the TPSD also converge demonstrating proportionality to $d_{par,50\%}$. This proportionality of TPSD with $d_{par,50\%}$ can be traced back to the proportionality of PSD to $d_{par,50\%}$ in Section 4.1. Because capillary trapping mainly occupies pore, the proportionality of PSD to $d_{par,50\%}$ also represents the proportionality of TPSD to $d_{par,50\%}$.

The TP cluster morphology can also be characterized through its volume ($V_{TP,cltr}$) and surface area ($A_{TP,cltr}$) as shown in **Figure 12**. Each point is the result obtained from averaging the $A_{TP,cltr}$ of all TP clusters in a log-scale range of $V_{TP,cltr}$. The examples of the TP clusters shape are also given in **Figure 12**. At smaller size, the markers from all porous media coincides one another. However, as the $V_{TP,cltr}$ increase, they branched out with TP clusters from smaller particle size exhibit larger $A_{TP,cltr}$. The point of the branch out is approximately at the size of $d_{pore,50\%}$. When the TP cluster size is smaller than d_{pore} of the porous media, it tends to occupy only one pore with a sphere-like shape. However, for larger cluster, it has to occupy more than one pore. As a result, it extends to the neighboring pores forming a channel of multi-pores TP cluster. These single-pore and multi-pores clusters can be modeled with a sphere and an elongated capsule model as follow:

$$A_{TP,cltr,SP} = 4.836 V_{TP,cltr}^{2/3} \quad (22)$$

$$A_{TP,cltr,MP} = 1.047 d_{pore,50\%}^{2} + \frac{4}{d_{pore,50\%}} V_{TP,cltr} \quad (23)$$

For the a_{TP}, it is given in **Figure 13a**. The line represents the cross-sectional average a_{TP} ($a_{TP,cs}$), whereas the marker represents the overall average a_{TP} ($a_{TP,ovr}$). **Figure 13a** shows that smaller particles generate larger a_{TP}. As described previously, smaller particles generate smaller PSD and TPSD. The single-pore TP clusters are sphere with smaller diameter, and the multi-pores TP clusters are elongated capsule with smaller diameter. As a result, the TP clusters generate higher a_{TP}. For the effect of particle size distribution range, similarly with PSD and TPSD, it is mainly controlled by the $d_{par,50\%}$. For the particle shape, PB generates the highest a_{TP}, whereas GB and TS generate similar a_{TP}. Given the PSD and TPSD, PB generates smaller pore and TP clusters, and thus higher a_{TP} was produced. GB and TS generate similar PSD but smaller than PB. Therefore, similar a_{TP} was produced between GB and TS but lower than PB. In addition, as shown in **Figure 9**, pores of PB is more irregular, and thus it generates TP clusters with more surface area. On the other hand, the pores of GB and TS are much less irregular than PB, and look relatively similar. Therefore, based on these two reasons, PB generate higher a_{TP}, whereas GB and TS generate similar a_{TP} but lower than PB. In addition, by normalizing a_{TP} with $d_{par,50\%}$, the a_{TP} converges into two lines (**Figure 13b**). All a_{TP} of PB converge into a single line,

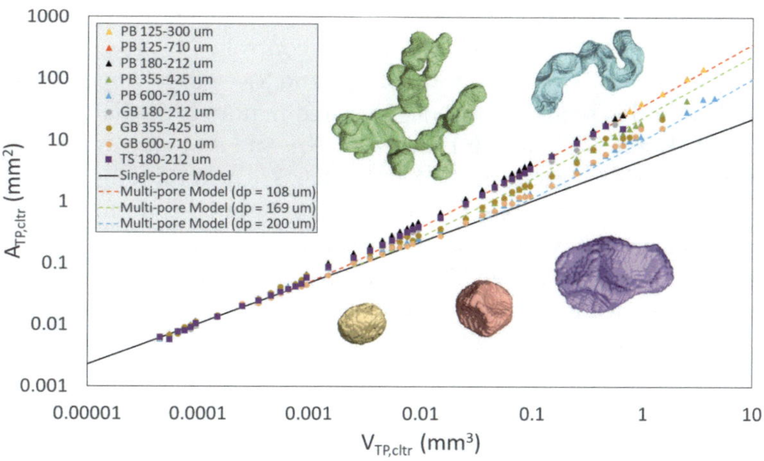

Figure 12.
The average $A_{TP,cltr}$ at the given $V_{TP,cltr}$ and examples TP clusters (Patmonoaji et al. [10]).

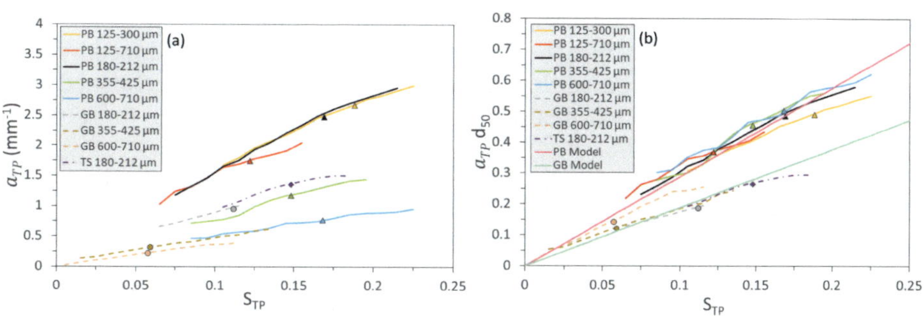

Figure 13.
(a) a_{TP} in all porous media and (b) $d_{par,50\%}$ normalized-a_{TP} with lines and markers represent $a_{TP,cs}$ and $a_{TP,ovr}$, respectively (taken with permission from Patmonoaji et al. [10]).

whereas TS and GB tend to converge into another line. Hence, two models were generated:

$$a_{TP} = 2.884 \frac{S_{TP}}{d_{par,50\%}} \tag{24}$$

$$a_{TP} = 1.889 \frac{S_{TP}}{d_{par,50\%}} \tag{25}$$

For the linearity between $\boldsymbol{a_{TP}}$ and $\boldsymbol{S_{TP}}$, it can be explained with these relations:

$$\frac{a_{TP}}{S_{TP}} = \Phi \frac{A_{TP}}{V_{TP}} \tag{26}$$

$$A_{TP} = \sum A_{TP,cltr,MP} \tag{27}$$

$$A_{TP} = \sum A_{TP,cltr,SP} = n_{TP,cltr} A_{TP,cltr,SP} \tag{28}$$

$$n_{TP,cltr} = \frac{V_{TP}}{V_{TP,cltr,SP}} \tag{29}$$

with $\boldsymbol{A_{TP}}$ as trapped phase interfacial area, $\boldsymbol{V_{TP}}$ as trapped phase volume, and $\boldsymbol{n_{TP,cltr}}$ as the number of TP clusters. When the trapped gas is mainly multi-pores TP clusters, which the function of $\boldsymbol{A_{TP,cltr}}$ is linear with $\boldsymbol{V_{TP,cltr}}$ (Eq. (23)), linear function between $\boldsymbol{a_{TP}}$ and $\boldsymbol{S_{TP}}$ will be generated. In contrast, if it consists of large number of single TP cluster, the relation will be controlled by $\boldsymbol{n_{TP,cltr}}$. Because $\boldsymbol{n_{TP,cltr}}$ is a function of $\boldsymbol{V_{TP}}$, the relation of $\boldsymbol{a_{TP}}$ and $\boldsymbol{S_{TP}}$ will be linear also.

As described in this section, the $\boldsymbol{a_{TP}}$ generated in porous media can be predicted by using Eqs. (24) and (25), which is based on $\boldsymbol{S_{TP}}$ and $\boldsymbol{d_{par,50\%}}$. Given these two parameters only, the $\boldsymbol{a_{TP}}$ can be predicted. Higher $\boldsymbol{S_{TP}}$ generates higher $\boldsymbol{a_{TP}}$. Smaller $\boldsymbol{d_{par,50\%}}$ also generates higher $\boldsymbol{a_{TP}}$. Porous media generated from more angular particles will also generates higher $\boldsymbol{a_{TP}}$. As described in Eq. (12), higher $\boldsymbol{a_{TP}}$ results in faster dissolution mass transfer. Therefore, porous media generated from angular and smaller particles leads to faster dissolution, especially at the condition of high $\boldsymbol{S_{TP}}$.

4.3 Flow bypassing

In general, flow bypassing can be divided into two categories. First is bypassing due to heterogeneity of porous media [32, 33], and second is bypassing induced by dissolution fingering in homogeneous porous media [34–36]. Both of them direct water flow to preferential flow path, resulting in non-uniform dissolution mass transfer. As a result, it will affect the $\boldsymbol{k}$. In this section, bypassing induced by dissolution fingering is examined through dissolution experiment performed in porous media from PB with various sizes and distributions [7] shown in **Table 2**. For the TP, N_2 gas was used.

In general, the mechanism of dissolution fingering development can be explained as follow. Initially, capillary trapping is relatively homogeneous, and thus the permeability tends to be homogeneous as well. However, a slight difference in permeability remains present. This slight difference generates preferential water flow, directing the water to flow to region with higher permeability. At this higher permeability region, water flow slightly faster, and thus the TP clusters dissolve slightly faster as well. As a consequence, it gradually increases the permeability of this slightly permeable region, enhancing the permeability differences gradually.

Size range (μm)	125–150	180–212	355–425	600–710	125–300	250–425	125–710
k category	low	low	high	high	low	high	high
λ (mm)	4.806	6.308	12.935	21.68	5.745	12.001	7.582

Table 2.
Characteristics of particle, porous medium, flow, and mass transfer inside each porous medium sample.

This results in a positive feedback, and thus the region of this preferential path grows with the dissolution time, resulting in the development of dissolution finger.

In short, the possibility of dissolution fingering development can be evaluated by measuring the approximated width of dissolution finger (λ) and compare it with the given size of porous medium diameter sample (L_y), which in this work is 10 mm. By using similar analysis for mineral dissolution instabilities based on linear stability analysis [37, 38], we can approximate the size of the dissolution finger by measuring the dimensionless growth rate ($\overline{\omega}$) using the calculation from Zhao et al. [36] as follow:

$$\overline{\omega} = \frac{\overline{p}'_{axf}}{(1+\beta)S_{N_2}}\left\{\frac{(1+\beta)\overline{p}'_{axf}}{1-\overline{\alpha}_L\overline{p}'_{axf}} + \left[(1+\beta) + \left(\frac{1}{1-\overline{\alpha}_L\overline{p}'_{axf}} - \overline{\alpha}_L\overline{m}\right)(1-\beta)\right]\sigma\right\} - \frac{\overline{p}'_{axf}}{(1+\beta)S_{N_2}}\left[\left(\overline{m} - \frac{\overline{p}'_{axf}}{1-\overline{\alpha}_L\overline{p}'_{axf}} - \overline{\alpha}_L\overline{m}^2\left(1-\overline{\alpha}_L\overline{p}'_{axf}\right)\right)\frac{(1-\beta)}{1-\overline{\alpha}_L\overline{p}'_{axf}}\right] \tag{30}$$

$$\sigma = \frac{\sqrt{\left(\frac{\overline{p}'_{axf}}{1-\overline{\alpha}_L\overline{p}'_{axf}}\right)^2 + \left(\frac{4\overline{m}^2\left(1-\overline{\alpha}_T\overline{p}'_{axf}\right)}{1-\overline{\alpha}_L\overline{p}'_{axf}}\right)} - \frac{\overline{p}'_{axf}}{1-\overline{\alpha}_L\overline{p}'_{axf}}}{2} \tag{31}$$

$$\beta = \left(\frac{1 - S_{N_2,0} - S_{wi}}{1 - S_{wi}}\right)^3 \tag{32}$$

$$\overline{p}'_{axf} = \frac{U}{\sqrt{\phi\tau D_{N_2}}}\sqrt{\frac{1}{\beta_0}} \tag{33}$$

$$\beta_0 = \frac{29.224\phi Re D_{N_2}}{{d_{50}}^2} \tag{34}$$

$$\overline{\alpha}_L = \frac{\alpha_L}{L^*} \tag{35}$$

$$\overline{\alpha}_T = \frac{\alpha_T}{L^*} \tag{36}$$

$$L^* = \sqrt{\tau D_{N_2} t^*} \tag{37}$$

$$t^* = \frac{\phi}{\beta_0} \tag{38}$$

with $\overline{m}$ as dimensionless wave number, S_{wi} as irreducible water saturation, $S_{N_2,0}$ as initial gas saturation, τ as tortuosity, α_L as longitudinal dispersivity, α_T as transversal dispersivity. By calculating the $\overline{\omega}$ at the given value of $\overline{m}$, the graph of $\overline{\omega}$ at the given $\overline{m}$ can be generated as shown in **Figure 14**. At the value of $\overline{\omega}$ lower than zero, the finger will stabilize, whereas at the value $\overline{\omega}$ larger than zero, the finger will develop. The peak value of $\overline{\omega}$ represents the value of critical dimensionless wave number ($\overline{m}_{critical}$), which the wave will grow fastest and most likely to be observed.

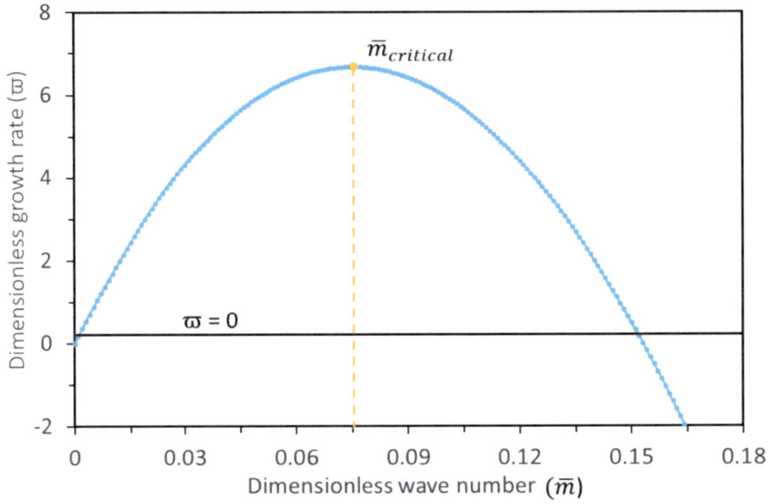

Figure 14.
The function of ϖ and $\overline{m}$ for the prediction of dissolution finger width calculated for PB 600–710 μm.

Therefore, by finding the $\overline{m}_{critical}$ in each experiment cases, λ in each experiment cases can be calculated as follow:

$$\lambda = \frac{2\pi L^*}{\overline{m}_{critical}} \tag{39}$$

The dissolution process inside PB 600–710 μm and PB 180–212 μm are given in **Figure 15a** as 3D images. In PB 600–710 μm, dissolution process occurs uniformly, whereas in PB 180–212 μm, dissolution occurs non-uniformly. The $\boldsymbol{k}$ throughout the dissolution process is given in **Figure 15b** with the function to $S_{TP,cs}$. Two groups of $\boldsymbol{k}$ are found. First group is higher $\boldsymbol{k}$, that occurs in PB 125–710 μm, PB 250–425 μm, PB 355–425 μm, and PB 600–710 μm. Second group is lower $\boldsymbol{k}$ that occur in PB 125–300 μm, PB 125–150 μm, and PB 180–212 μm. From the predicted λ, the first group belong to the value of λ larger than L_y, whereas the second group belong to the value of λ smaller than L_y. When the size of λ is smaller than L_y, dissolution fingers could fully develop, resulting in bypassing and lower $\boldsymbol{k}$. This trend occurs except in PB 125–710 μm that was influenced by the particle size distribution. The classification of $\boldsymbol{k}$ and predicted λ is given in **Table 2**.

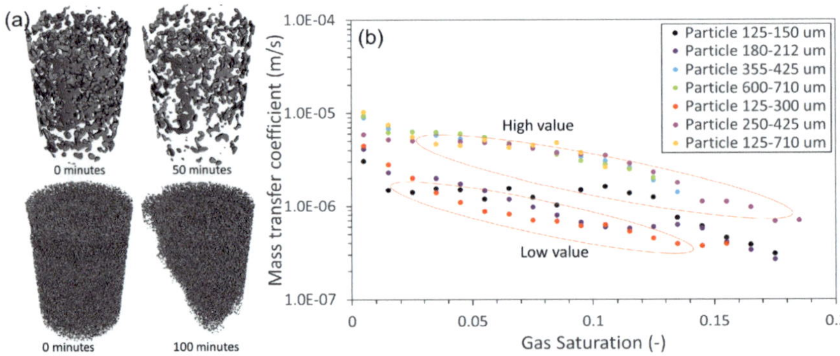

Figure 15.
*(a) Dissolution progress of (top) particle 600-710 μm and (bottom) particle 180-212 μm and (b) **k** at the given saturation in all porous media (modified from Patmonoaji et al. [7]).*

As explained in this section, in the presence of bypassing induced by dissolution fingering, $\boldsymbol{k}$ is lower due to non-uniform dissolution process. Because this is also occur in homogeneous porous media, and the limiting condition is just the ratio of the container to the predicted size of dissolution fingering, it is expected to occurs in field-scale of hydrogeology processes.

4.4 Trapped phase properties

Another parameter that affect the dissolution mass transfer rate is $\boldsymbol{\xi}$, which is the ratio of $\boldsymbol{C_{i,sol}}$ and $\boldsymbol{\rho_i}$. $\boldsymbol{\rho_i}$ affects the decrease rate in volume of dissolution process, and $\boldsymbol{C_{i,sol}}$ serves as the driving force of the mass transfer process. Higher $\boldsymbol{\xi}$ results in faster dissolution mass transfer process. As comparison, $\boldsymbol{\xi}$ varies for about 10^{-9}–10^{-3} in mineral dissolution, 10^{-4}–10^{-3} in NAPL dissolution, and 10^{-3}–10^{-1} in gas dissolution [36]. Timewise, this implicates that the dissolution of minerals could take months and years, whereas dissolution of NAPL could take days to weeks. In contrast, dissolution of gases could only take minutes to hours.

Dissolution in porous media is not only mass transfer process, but also solute transport. When mass transfer occurs from TP, the dissolve solute need to be transported away from the surrounding TP by the FP. Therefore, high value in $\boldsymbol{\xi}$ could generate problems in the balance of mass transfer and solute transport.

To compare the effect $\boldsymbol{\xi}$, CO_2 and N_2 gases were used. The $\boldsymbol{\rho_i}$ of N_2 and CO_2 gases are 1.842 and 1.165 kg/m^3, whereas the $\boldsymbol{C_{i,sol}}$ are 1.449 and 0.0175 kg/m^3, resulting in $\boldsymbol{\xi}$ of 0.787 and 0.015, respectively. The differences in $\boldsymbol{\xi}$ is more than fifty times. The experiments were performed with the same porous media and FP velocity, resulting in the same $\boldsymbol{Re}$.

The axial average TP saturation throughout the dissolution process is given in **Figure 16**. The dissolution of N_2, and CO_2 gases behave differently. In case of N_2 gas, as the time proceeds, $\boldsymbol{S_{TP,cs}}$ decrease from the initial value to zero. For CO_2, on the other hand, the $\boldsymbol{S_{TP,cs}}$ does not dissolve completely but stop at the value about

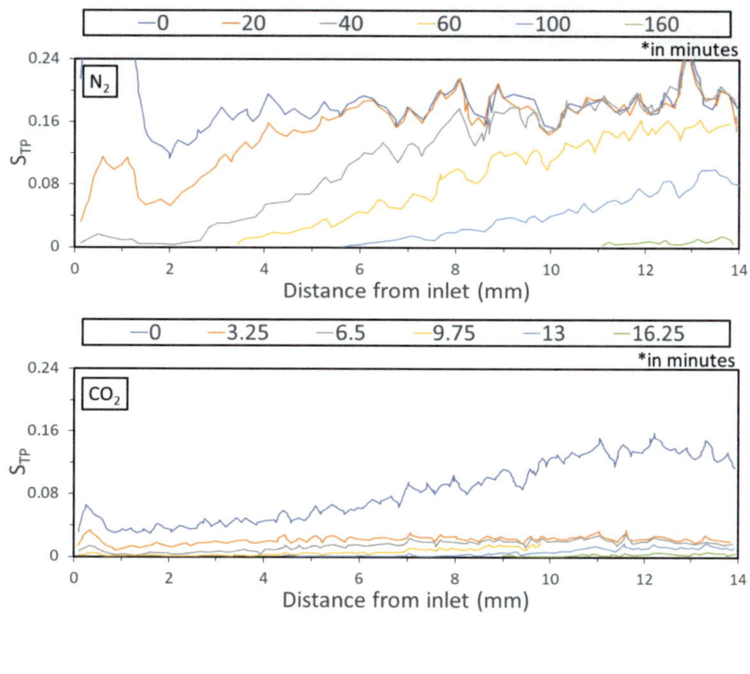

Figure 16.
$S_{TP,cs}$ of N_2 and CO_2 gas during dissolution process under Re of 0.04.

0.03 before it dissolves completely, resulting in two dissolution stages. First dissolution stage occurs as the dissolution of $S_{TP,cs}$ to about 0.03. Second dissolution stage occurs as the complete dissolution of the remaining $S_{TP,cs}$. By varying $\boldsymbol{Re}$, the same dissolution pattern of CO_2 is still observed [9].

By observing the dissolution process from the 3D images of segmented TP clusters as shown in **Figure 17**, the different dissolution behavior can be observed in more detail. In case of N_2, TP clusters dissolve completely from multi-pores size. However, in case of CO_2, at first, the multi-pores TP clusters dissolve to single-pore TP clusters. Afterwards, second stage occurs, dissolving the remaining clusters.

For the $\boldsymbol{k}$, additionally, there is a significant different between CO_2 and N_2 gas as shown in **Figure 18**. Given the same $\boldsymbol{Re}$, the $\boldsymbol{k}$ of CO_2 is more than one order of magnitude lower than N_2 gas. In addition, $\boldsymbol{k}$ shows a dipping point in around 0.03, which is consistent with the two stages dissolution of CO_2. When the $S_{TP,cs}$ reaches 0.03, the dissolution process slowed down, and thus $\boldsymbol{k}$ decreases. Then second dissolution stages occurs, dissolving the rest $S_{TP,cs}$, resulting in the increase of $\boldsymbol{k}$.

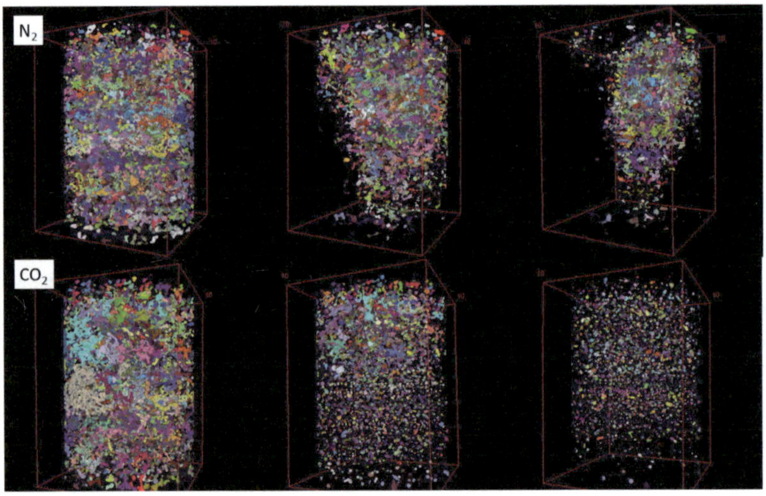

Figure 17.
TP cluster of N2 and CO_2 observed at different dissolution time. Different colors indicate different clusters (modified from Patmonoaji et al. [9]).

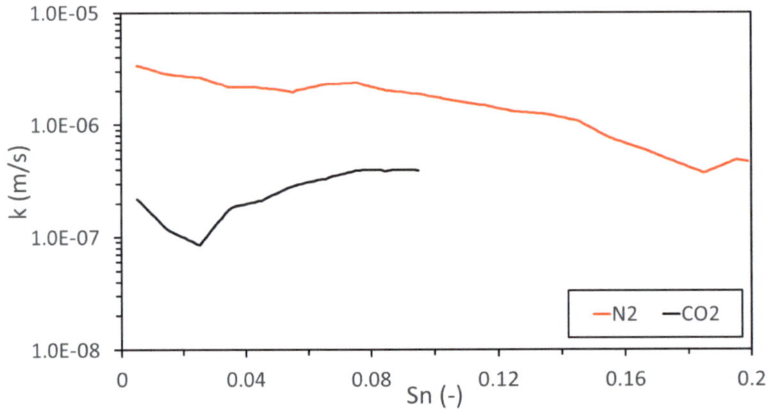

Figure 18.
Average $\boldsymbol{k}$ of all experiments with different gases and $\boldsymbol{Re}$.

Compared with other gases, CO_2 exhibit about fifty times larger. When TP cluster dissolves, the solute should be transported away by fresh incoming FP. If the solute is not transported, the solute concentration surrounding the TP cluster accumulates, slowing down the mass transfer. During the dissolution of CO_2, this rapid dissolution is believed to generate area with high CO_2 solute concentration due to much faster mass transfer rate than advection rate. In addition, as the trapped gas dissolves rapidly, it quickly retracts to the center of pores, increasing the difficulty of the solute to be removed by FP. As the dissolution occurs rapidly, the incoming FP cannot compete with the rapid dissolution of TP. As a result, two dissolution stages occur. First stage occurs before the accumulation of solute, whereas the second stage occurs after the accumulated solute is advected away from the surrounding of TP.

As discussed in this section, although higher ξ results in higher dissolution mass transfer rate, when it becomes too high, it could result in lower k. This condition can be represented by two dissolution stages that occurs due to competition between mass transfer rate and advection rate.

4.5 Flowing phase velocity

To investigate the effect of FP velocity, the experiments were also performed under different FP velocity (Re between 0.0016 and 0.16) with N_2 and CO_2 gases. They were performed with porous media from PB 250–425 μm to avoid bypassing and focus the discussion to the effect of FP velocity. By using the Sh and Re, the mass transfer can be modeled with Eq. (6) as follow (**Figure 19**):

$$Sh = 12.142\, Re \tag{40}$$

$$Sh = 0.386\, Re^{0.685} \tag{41}$$

The Eq. (40) is for N_2 gas, whereas Eq. (41) is for CO_2 gas. The difference is due to the effect of high ξ in CO_2, resulting in two dissolution stages. Because the k is also lower, the non-dimensional number model result in lower Sh than N_2 gas.

In this section, the effect of FP velocity was shown. With the increase in FP velocity, k increase. This relation can also be represented in non-dimensional number, as represented by Re and Sh in Eqs. (40) and (41). The effect of high ξ is also observable as lower Sh due to lower k.

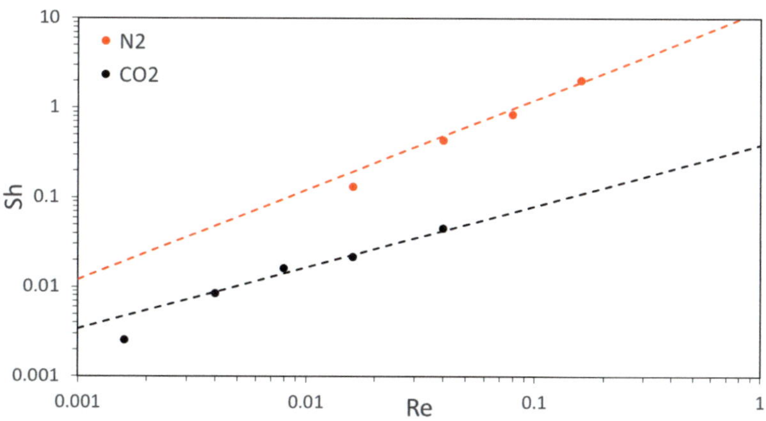

Figure 19.
*Comparison of predicted-**Sh** and measured-**Sh** from the model.*

5. Conclusion

Throughout this work, the physical mechanism and mathematical model of dissolution mass transfer of trapped phase in porous media was described. Various condition affecting it were also discussed from experiment data obtained from micro-tomography and a series of image processing techniques. The effects of these conditions were represented in three parameters: $\boldsymbol{A}$, $\boldsymbol{\xi}$, and $\boldsymbol{k}$.

In case of porous media characteristics, the characteristics of porous media generated from various particle shape, size, and distribution were investigated. Smaller $\boldsymbol{d_{par,50\%}}$ results in higher $\boldsymbol{a_{TP}}$. Porous media generated from more angular particles will also generates higher $\boldsymbol{a_{TP}}$. In addition, $\boldsymbol{S_{TP}}$ linearly effect the amount of $\boldsymbol{a_{TP}}$. Therefore, porous media generated from angular and smaller particles leads to faster dissolution, especially at the condition of high $\boldsymbol{S_{TP}}$.

Bypassing induced by dissolution fingering was also described. When the predicted λ is smaller than $\boldsymbol{L_y}$, dissolution finger can fully develop, resulting in bypassing and decrease in $\boldsymbol{k}$. Because this occurs in homogeneous porous media, it is expected to occurs in field-scale of hydrogeology processes.

In addition, the effect of $\boldsymbol{\xi}$ was also investigated. If the $\boldsymbol{\xi}$ of the TP is high, different dissolution process, which is represented by two stage dissolutions, occurs, leading to decrease in $\boldsymbol{k}$. The competition between mass transfer and advection rate is believed to be the reason of this different dissolution process.

Eventually, the effect of FP velocity based on $\boldsymbol{Sh}$ and $\boldsymbol{Re}$ was formulated. With the increase of FP velocity, $\boldsymbol{k}$ increase as represented by the increase in $\boldsymbol{Sh}$ at the given $\boldsymbol{Re}$. In case of CO_2, the $\boldsymbol{Sh}$ also increase with $\boldsymbol{Re}$, but it is still lower than N_2. This is due to the effect two stage dissolution observed in CO_2, resulting in lower $\boldsymbol{k}$.

Acknowledgements

This work was supported by JSPS KAKENHI with grant numbers 17H00790 and 20 J14975.

Author details

Anindityo Patmonoaji*, Yingxue Hu, Chunwei Zhang and Tetsuya Suekane
Tokyo Institute of Technology, Japan

*Address all correspondence to: patmonoaji.a.aa@m.titech.ac.jp

References

[1] Miller CT, Poirier-Mcneill MM, Mayer AS. Dissolution of trapped nonaqueous phase liquids: mass transfer characteristics. Water Resour Res. 1990; 26(11):2783–2796.

[2] Powers SE, Abriola LM, Weber WJ. An experimental investigation of nonaqueous phase liquid dissolution in saturated subsurface systems: Steady state mass transfer rates. Water Resour Res. 1992;28(10):2691–2705.

[3] Imhoff PT, Jaffé PR, Pinder GF. An experimental study of complete dissolution of a nonaqueous phase liquid in saturated porous media. Water Resour Res. 1994;30(2):307–320.

[4] Donaldson JH, Istok JD, Humphrey MD, O'Reilly KT, Hawelka CA, Mohr DH. Development and testing of a kinetic model for oxygen transport in porous media in the presence of trapped gas. Ground Water. 1997;35(2):270–279.

[5] Geistlinger H, Beckmann A, Lazik D. Mass transfer between a multicomponent trapped gas phase and a mobile water phase: Experiment and theory. Water Resour Res. 2005;41(11): 1–15.

[6] Iglauer S. Dissolution trapping of carbon dioxide in reservoir formation brine - a carbon stoage mechanism. In: Nakajima H, editor. Mass Transfer - Advanced Aspects. IntechOpen; 2011. p. 233–262.

[7] Patmonoaji A, Hu Y, Zhang C, Tsuji K, Suekane T. Investigation on the effect of particle size in dissolution mass transfer inside porous media with micro-tomography. AIP Conf Proc. 2020;2248(July).

[8] Chang C, Zhou Q, Guo J, Yu Q. Supercritical CO2 dissolution and mass transfer in low-permeability sandstone: Effect of concentration difference in water-flood experiments. Int J Greenh Gas Control. 2014;28:328–342.

[9] Patmonoaji A, Suekane T. Investigation of CO2 dissolution via mass transfer inside a porous medium. Adv Water Resour. 2017;110(May):97–106.

[10] Patmonoaji A, Tsuji K, Muharrik M, Suekane T. Micro-tomographic analyses of specific interfacial area inside unconsolidated porous media with differing particle characteristics from microscopic to macroscopic scale. J Colloid Interface Sci. 2018;532:614–621.

[11] Patmonoaji A, Tsuji K, Suekane T. Pore-throat characterization of unconsolidated porous media using watershed-segmentation algorithm. Powder Technol. 2020;362:635–644.

[12] Patmonoaji A, Muharrik M, Hu Y, Zhang C, Suekane T. Three-dimensional fingering structures in immiscible flow at the crossover from viscous to capillary fingering. Int J Multiph Flow. 2020;122.

[13] Bird RB, Stewart WE. Transport Phenomena. Second. Wiley; 2006. 928 p.

[14] Cussler EL. Diffusion mass transfer in fluid systems. Third. New York: Cambridge Univ. Press,; 2009.

[15] Cengel YA. Heat Transfer: A Practical Approach. Second. Mc Graw-Hill. McGraw-Hill; 2002. 896 p.

[16] Graham T. On the Law of the Diffusion of Gases. Philos Mag. 1833;2.

[17] Fourier J. Théorie analytique de la chaleur. Paris; 1822.

[18] Fick A. On Liquid Diffusion. 1855; 30–39.

[19] Nernst W. Theorie der Reaktionsgeschwindigkeit in heterogenen Systemen. Zeitschrift fur Phys chemie. 1904;47:52.

[20] Higbie R. The rate of absorption of a pure gas into still liquid during short periods of exposure. Trans Am Inst Chem Eng. 1935;31:365.

[21] Danckwerts P V. Gas absorption accompanied by a two-step chemical reaction. AIChE J. 1955;1:456–463.

[22] Darcy H. Les fontaines publiques de la ville de Dijon. V. Dalmont; 1856. 647 p.

[23] Nutting PG. Physical analysis of oil sands. Am Assoc Pet Geol Bull. 1930;14 (10):1337–1349.

[24] Wyckoff RD, Botset HG, Muskat M, Reed DW. The measurement of the permeability of porous media for homogeneous fluids. Rev Sci Instrum. 1933;4(7):394–405.

[25] Abdon Atangana. Fractional Operators with Constant and Variable Order with Application to Geo-Hydrology. Academic Press; 2017. 414 p.

[26] Bear J. Dynamics of fluids in porous media. Revised. Mineola, New York: Dover; 1988.

[27] Sahimi M. Flow and transport in porous media and fractured rock. second. Wiley-VCH; 2011.

[28] Dullien FAL. Porous Media: Fluid Transport and Pore Structure. Second. New York: Elsevier; 1992.

[29] Blunt MJ. Multiphase flow in permeable media : a pore scale perspective. London: Cambridge University Press; 2017.

[30] Morrow NR, Songkran B. Effect of viscous and buoyancy forces on nonwetting phase trapping in porous media. Surf Phenom Enhanc Oil Recover. 1981;387–411.

[31] Wardlaw NV, Mckellar M. Oil blob populations and mobilization of trapped oil in unconsolidated packs. Can J Chem Eng. 1985;63(4):525–532.

[32] Patmonoaji A, Zhang Y, Xue Z, Park H, Suekane T. Experimental and numerical simulation of supercritical CO 2 microbubble injection into a brine-saturated porous medium. Int J Greenh Gas Control. 2019;91(September): 102830.

[33] Soerens TS, Sabatini DA, Harwell JH. Effects of flow bypassing and nonuniform NAPL distribution on the mass transfer characteristics of NAPL dissolution. Water Resour Res. 1998;34(7):1657–1673.

[34] Imhoff PT, Thyrum GP, Miller CT. Dissolution fingering during the solubilization of nonaqueous phase liquids in saturated porous media 2. Experimental observations. Water Resour Res. 1996;32(7):1929–1942.

[35] Imhoff PT, Miller CT. Dissolution fingering during the solubilization of nonaqueous phase liquids in saturated porous media: 1. Model predictions. Water Resour Res. 1996;32(7):1919–1928.

[36] Zhao C. Physical and chemical dissolution front instability in porous media. Springer; 2014. 354 p.

[37] Chadam J, Hoff D, Merino E, Ortoleva P, Sen A. Reactive infiltration instabilities. IMA J Appl Math (Institute Math Its Appl. 1986;36(3):207–21.

[38] Chadam J, Ortoleva P. Morphological instabilities in physico-chemical systems. Earth Sci Rev. 1990; 29(1–4):175–181.

Section 4

Heat and Mass Transfer in Porous Media

Chapter 5

Heat Transfer Enhancement Using Unidirectional Porous Media under High Heat Flux Conditions

Kazuhisa Yuki

Abstract

In this chapter, new heat transfer enhancement technologies with unidirectional porous metal called "EVAPORON" and "Lotus' Breathing" are introduced to remove and manage heat from high heat flux equipment. The unidirectional porous metals introduced here can be easily fabricated by unique techniques such as mold casting technique, explosive welding technique, and 3D printing technique. First of all, many kinds of porous media, which have been introduced by the author so far as a heat transfer promoter, are compared each other to clarify what kind of porous metal is more suitable for high heat flux removal and cooling by focusing on the permeability and the effective thermal conductivity. For the practical use of the unidirectional porous copper with high permeability and high thermal conductivity, at first, heat transfer performance of two-phase flow cooling using a heat removal device called "EVAPORON" is reviewed aiming at extremely high heat flux removal beyond 10 MW/m^2. We have been proposing this device with the unidirectional porous copper fabricated by 3D printing technique as the heat sink of a nuclear fusion divertor and a continuous casting mold. Second, two-phase immersion cooling technique called "Lotus' Breathing" utilizing "Breathing Phenomenon" is introduced targeting at thermal management of various electronics such as power electronics and high performance computers. The level of the heat flux is 0.1 MW/m^2 to 5 MW/m^2. In addition, as the other heat transfer enhancing technology with unidirectional porous metals, unidirectional porous copper pipes fabricated by explosive welding technique are also introduced for heat transfer enhancement of single-phase flow.

Keywords: unidirectional porous metal, cooling, high heat flux, permeability, effective thermal conductivity, heat transfer, phase change, EVAPORON, lotus' breathing

1. Introduction

Porous media are defined as "materials containing very large numbers of pores with various sizes." Owing to their characteristic large surface area, the porous media are used in various fields such as chemical plants, architecture, agriculture, environment, medical care, biology etc. In heat transfer engineering, which is one of this book's topics, heat pipes [1] and vapor chambers [2] that utilize independent liquid supply based on the capillarity in a porous wick play an active part in the thermal management of electronic devices. For instance, thin heat pipes of <1 mm in thickness are even used in smartphones whose usage has witnessed an exponential increase [3]. In addition, in recent years, to cool electronic devices, two-phase immersion cooling with saturated pool boiling heat transfer has attracted attention for increasing the heat generation density and reducing the costs of thermal management. In particular, great numbers of studies have focused on improving the critical heat flux by loading a porous medium with a functional shape and structure onto the cooling surface, in addition to enhance the boiling heat transfer. It is recommended to refer to the review studies of Yuki [4], Mori [5, 6], and Kandlikar et al. [7]; however, there have been multiple studies that attempted to separate liquid supply and vapor discharge, such as the honeycomb porous plate of Mori et al. [8], the porous layers having a non-uniform spatial structure [9–11], and porous meshes provided with hydrophilic and hydrophobic functions [12, 13]. Note that the results of Bai et al. [14] and Yuki et al. [15] in which the critical heat flux of the saturated pool boiling of water in an atmospheric pressure environment exceeded 5 MW/m^2, thus demonstrating the high potential of the porous media as a latent heat transport promoter. On the other hand, also for a single-phase flow, there have been a large number of attempts to utilize the porous media as a heat transfer promoter. The examples include heat transfer enhancement by metal foams [16, 17], utilization as a radiation converter [18, 19], and promotion of heat transfer using sphere-packed tubes [20, 21].

Focusing on high potential of the porous media as the heat transfer promoter as mentioned above, multiple studies have introduced porous metals for heat removal from high-heat flux equipment of a level of 10 MW/m^2. For example, the divertor of a nuclear fusion reactor currently under research and development is exposed to a steady heat load of approximately 10 MW/m^2 due to the inflow of α particles generated by nuclear fusion reaction. Sharafat et al. [22, 23] introduced metal foams as a heat transfer promoter of He gas flow pressurized to 40 MPa and demonstrated high cooling performance of >10 MW/m^2. Furthermore, Joshi et al. proposed a cooling technology that makes use of the phase change of coolant in a pin-fins microchannel for high heat generation density electronic devices exceeding 10 MW/m^2 [24–26]. Moreover, he attempted a technology that assists heat transfer in the promotion of evaporation in a nanoporous film via a gas impinging jet [27]. Regarding additional applications of the porous media under high heat flux conditions, refer to the review article by Smakulski et al. [28]. Obviously, in the thermal management of these high heat flux devices, in addition to securing the cooling performance, it is necessary to consider both economical efficiency and maintainability; therefore, thermal management with low flow rate and low pumping power is indispensable.

As an efficient heat removal technology at a level of 10 MW/m^2, Toda and Yuki proposed a cooling device known as EVAPORON (Evaporated Fluid Porous-Thermodevice) to take maximum advantage of the latent heat of vaporization of the coolant [29, 30]. EVAPORON can remove extremely high heat flux by evaporating the cooling liquid that is fed inside the porous medium jointed to the heat transfer surface against a heat flow, using the vast surface area with microchannels of the porous media. Because rather than relying on capillarity such as in the heat pipes, a minimum amount of liquid corresponding to the heat flux level is gently pumped into the porous medium for cooling, a significant reduction in pumping power can then be expected. Furthermore, we verified that the important factor of this technological breakthrough is the quick discharge of considerable amounts of vapor generated in the porous medium [31]. Subsequently, the authors proposed EVAPORON-2 [32, 33] loaded with subchannels for vapor discharge within sintered copper particles and succeeded in the heat removal of over 20 MW/m^2. Targeting at a cooling performance exceeding 10 MW/m^2 under much lower pumping power conditions, furthermore, the EVAPORON-3 equipped with a vapor–liquid separator on the upper part of the porous medium was also proposed [34].

Figure 1.
Various kinds of porous media.

To summarize, when applying the porous media in the two-phase cooling technology for high heat flux environment, the most important factor is to actively promote the vapor discharge that was generated in the porous medium. Moreover, it is necessary to optimize the porous structure to maximize the advantage of the latent heat of vaporization of the coolant. On the other hand, there are various porous media such as foams, open cells, sintered particle/fiber, meshes etc. (see **Figure 1**), even if all of these are simply called porous metals. Generally, it is impossible to generalize the thermophysical properties and mechanical properties of the porous media with these different structures, thus referring only on their porosity and pore size. However, to promote phase change heat transfer in the porous medium, it is an undisputed fact that a porous medium with high permeability, which enables rapid discharge of the vapor generated, should be introduced in which its thermal conductivity is also increased to expand the effective heat transfer area. In general, there is a trade-off relationship between increasing the effective thermal conductivity and reducing the flow resistance of the vapor flow (increasing the permeability). Thus, the porous media to be introduced should be selected based on the heat flux level and coolant used. In this chapter, the effective thermal conductivity and absolute permeability for various porous media are first discussed. Subsequently, the effectiveness of unidirectional porous coppers [35] proposed by the author is quantitatively evaluated as a cooling technique on a level of 10 MW/m^2. Finally, two our innovative cooling technologies are introduced; "EVAPORON-4", which is the newest EVAPORON that combines the unidirectional porous copper and a grooved heat transfer surface and "Lotus' Breathing", which makes use of two-phase immersion cooling in a saturated pool boiling environment, based on spontaneous liquid supply effect called "Breathing phenomenon" proposed by the author for the first time all over the world.

2. Thermal conductivity and permeability of various porous media

Figure 2 shows the porosity and pore size distributions of sintered-metal-particles/fibers and foamed metals the author used to date. For example, regarding the thermal conductivity and the permeability of isotropic porous media with a simple structure, their modeling is frequently performed based on the porosity and the pore size for each porous medium. As shown in **Figure 2**, each porous medium has its characteristic porosity and pore diameter. For instance, most foamed metals and open cells have a high porosity of >90%. As far as the author knows, although those with a microscale pore diameter are not commercially available, Unno et al. developed the porous media with high porosity and micropores and demonstrated that it is effective for promoting evaporative heat transfer as per the capillary limit theory [36]. Multiple sintered-metal-particles have a porosity of 30%–50% because

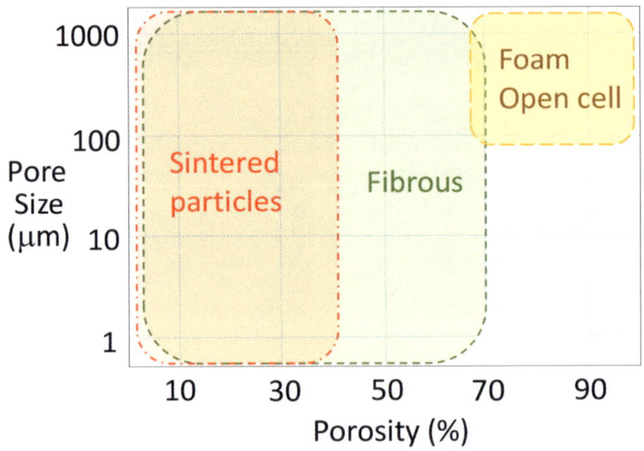

Figure 2.
Porosity and pore size of porous media.

of the producing process and the packing structure; moreover, the pore diameter varies from the order of micron to millimeter. For sintered-metal-fibers (a kind of fibrous porous media), the porosity can be regulated over a wide range by the extent of pressing; however, it is difficult to control the orientation of the fiber. We will now discuss the effective thermal conductivity and permeability of these typical porous media from the view point of phase-change enhancement of coolant.

2.1 Effective thermal conductivity of porous media

To discuss the effective heat transfer area, it is important to examine the effective thermal conductivity of the porous media. The prediction formulae for various porous structures have been developed to date; however, in general, the effective thermal conductivity k_{eff} of the porous media can be discussed based on the following parallel model weighted by the porosity.

$$k_{eff} = \varepsilon k_f + (1-\varepsilon) k_s \tag{1}$$

where k_f and k_s are the thermal conductivity of the fluid phase in the pore and that of the porous solid phase, respectively. Note that both k_f and k_s do not express the thermal conductivity of the material itself but must be taken into account the tortuosity, which expresses the porous structure, and the thermal resistance affected by the degree of sintering, if speaking of a sintered metal. However, Eq. (1) clearly shows that the effective thermal conductivity of the porous media mostly depends on the porosity. For example, for copper-foamed metal with high porosity, if the fluid in the pore is air and the porosity is 0.9, the effective thermal conductivity is estimated to be somewhat <40 W/m/K; however, when estimated from the formula of Boomsma et al. [37], it is ~1 W/m/K. In fact, because the effective thermal conductivity decreases further, depending on the manufacturing method of foamed metals, the expansion of the heat transfer area by the fin effect cannot be expected, particularly for liquid cooling or two-phase immersion cooling. Therefore, the foamed metals and the open cells are often used for gas flows as mechanical dispersion promoters and turbulent promoters. On the other hand, in terms of sintered-metal-particles with a comparatively low porosity, the porous media with the porosity of <0.3 can be produced depending on the degree of sintering and pressing, so that high effective thermal conductivity can be expected. Furthermore, depending on the extent of sintering, the neck structure formed between the particles considerably affects the effective thermal conductivity. As for the fibrous porous media, although the sintered-metal-fibers have high thermal conductivity in the fiber direction, the effective thermal conductivity is assumed to be almost the same as that of the sintered-metal-particles in the direction perpendicularly to the fiber direction.

Herein, as shown in **Figure 3**, we performed heat conduction simulation to estimate the effective thermal conductivity of the sintered-copper-particle, in which a cubic sintered-copper-particle with a side length of 5 mm is sandwiched between copper square rods of 5 × 5 × 50 mm^3. At the lower end surface, the heat flux of 0.5 MW/m^2 is exposed and a constant temperature of 100 °C is given to the upper end surface. Adiabatic conditions are attributed to the side surfaces of the rods and sintered-particles. The effective thermal conductivity of the sintered-copper-particle is evaluated using the Fourier law from the difference in the average temperature at the two interfaces between the copper square rod and the sintered particle. The particle size is 1.0 mm and packed structure of particles is a simple cubic structure. To reproduce the neck structure formed between the particles during sintering, a cylinder with a diameter of d is virtually installed around the contact point between the particles, i.e., the diameter of the cylinder is the neck structure at the time of sintering in a pseudo presentation, and d is determined with the contact angle θ as a parameter. In this calculation, contact angles are 5°, 10°, and 20°. The same is valid for contact between the particles and the end surface of the rods. The porosity at a contact angle of 0° is 0.48. Assuming pure copper, the solid phase has a thermal conductivity of 398 W/m/K; moreover, the air in the pore has a thermal conductivity of 0.026 W/m/K. The heat conduction simulation is performed using Stream v13 developed by Cradle, which adopts the finite volume method (FVM); for comparison, the calculation using the finite element method (FEM) is performed at the same time. Because these simulations do not apply any special model for the discretization of the equation of heat conduction, the effective thermal conductivity of the sintered-particle can be evaluated with a higher accuracy than with experiments. **Figure 4** shows the results of the simulation. The effective thermal conductivity at contact angles of 5°, 10°, and 20°is 7.8, 19.2, and 45.9 W/m/K, respectively. Obviously, a higher degree of sintering involves higher effective thermal conductivity. However, compared with the bulk thermal conductivity of pure copper, it is reduced to 2% – 4% of the thermal conductivity of the Cu

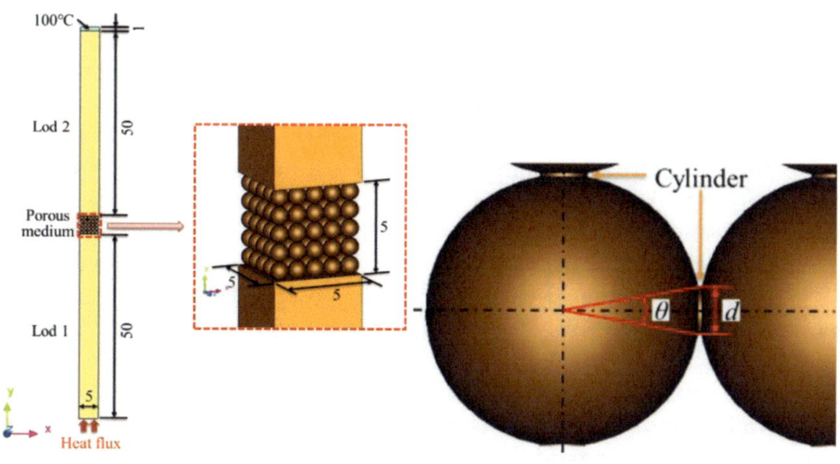

Figure 3.
Calculation of effective thermal conductivity for particle-sintered porous medium.

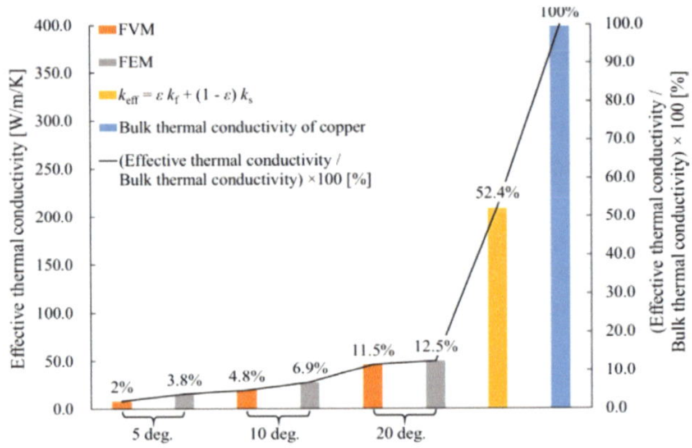

Figure 4.
Effective thermal conductivity of particle-sintered porous medium.

material at a contact angle of 5°; even if the contact angle is 20°, which is the highest degree of sintering, the effective thermal conductivity is reduced to ~12% of the Cu material. The difference from the effective thermal conductivity represented by the parallel model of Eq. (1) is apparent. The abovementioned results verify the necessity to carefully consider the fin effect even when utilizing the sintered-copper-particles due to its low thermal conductivity.

On the other hand, to maximize the effective thermal conductivity of the porous media, a porous medium to which the parallel model of Eq. (1) can be applied is desirable. This indicates that the introduction of a porous medium with unidirectional pore structure is ideal. Its thermal conductivity differs between the direction of the pores ($k_{eff//}$) and the direction perpendicular thereto ($k_{eff\perp}$); they can be predicted from the following equations by Ogushi et al. [38].

$$k_{eff\parallel} = (1-\varepsilon)k_s \tag{2}$$

$$k_{eff\perp} = \frac{(\beta+1)+\varepsilon(\beta-1)}{(\beta+1)-\varepsilon(\beta-1)}k_s \tag{3}$$

The abovementioned equation is driven by adapting the Behrens' thermal conductivity model for composite material to the unidirectional porous media. Because the fluid in the pore is assumed to be a gas and its thermal conductivity can be ignored, Eq. (2) corresponds to the parallel model. β is the thermal conductivity ratio. At the same porosity of 0.48 as in the sintered-particle, the effective thermal conductivity in the pore direction is 220 W/m/K, which is more than five times higher than that of the sintered-particle. When developing cooling devices for a high heat flux equipment, if the pore direction of the unidirectional porous media agrees with the heat flow direction, the effective heat transfer area potential of the porous media can be maximized. With regard to the effective thermal conductivity in the direction perpendicular to the pores, the thermal conductivity simulation as shown in **Figure 3** is also performed on the unidirectional porous media as shown in **Figure 5** whereby the effective thermal conductivity is predicted more accurately. In this simulation, the pores are arranged in two patterns, i.e., a square array and a staggered array, and the porosity is adjusted while changing the pore diameter from 0.1 to 1.9 mm without changing the pore position. As understood from **Figure 6**, the unidirectional porous media exhibits a similarly high thermal conductivity even in the direction perpendicular to the pores because the effective thermal conductivity with the porosity of 0.5 exceeds

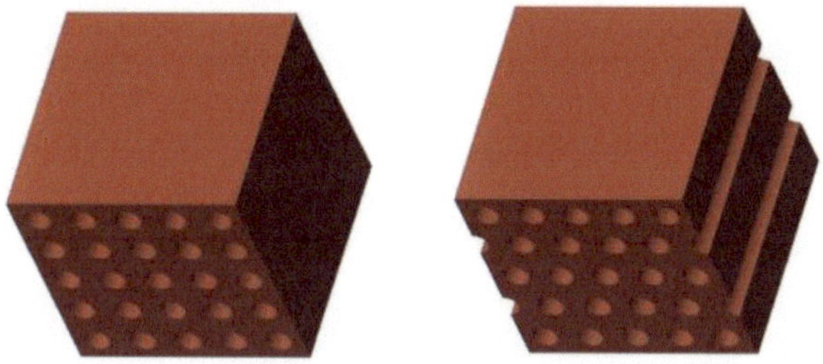

Figure 5.
Square array model and staggered array model for uni-directional porous media.

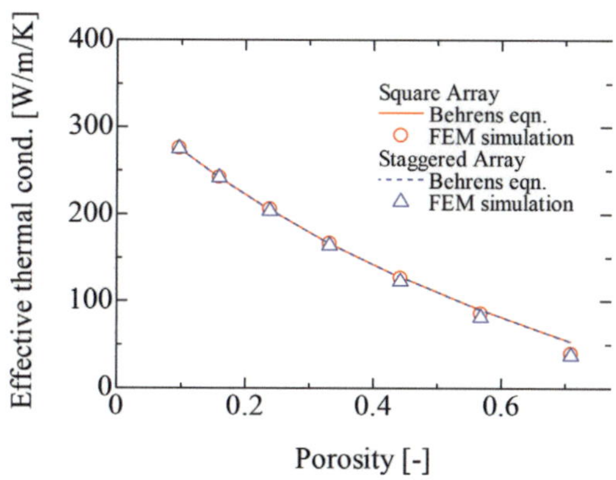

Figure 6.
Effective thermal conductivity $k_{eff\perp}$ of uni-directional porous media.

100 W/m/K. Moreover, the higher the porosity, the larger the deviation of the effective thermal conductivity from the model of Chiba et al. For example, if the porosity is 0.7, the error is ~35% for the square array and ~ 42% for the staggered array. By grasping the effective thermal conductivity, it becomes possible to discuss the fin efficiency of the porous media. Chiba et al. developed a highly accurate model formula [39], which enables the easy evaluation of the criteria for introducing the unidirectional porous media.

2.2 Permeability of porous media

As mentioned above, to effectively use the latent heat of vaporization in a high heat flux environment, a large amount of vapor generated in the porous medium must be quickly discharged to the outside. In other words, it is indispensable to take a mean, including introducing the porous media with high permeability or modifying the spatial structure while considering the improvement in effective thermal conductivity discussed in the previous section. Here, for reference, a simple comparison of the absolute permeability K of the sintered-particle and the foamed metal from the Kozeny–Carman Eq. $K = d^2\varepsilon^3/180/(1\text{-}\varepsilon)^2$ demonstrates that the permeability of the foamed metals is by two orders of magnitude higher than that of the sintered-particles, which clearly shows outstanding fluid fluidity and vapor discharge performance. However, as we have already understood from the discussion on the effective thermal conductivity, if the foamed metal is used as the phase change promoter, the phase change within the porous medium cannot be expected because of the low effective thermal conductivity, i.e., boiling/evaporation is expected to occur only on the heat transfer surface or in its immediate neighborhood. As for the sintered-particles, complicated microchannels with repeated expanded and contracted portions can be considered particularly suitable for the phase change because a thin liquid film can be formed at the contracted portion. Because of an extremely large pressure loss for the sintered-particles compared to that for the foamed metals, however, it is difficult to smoothly discharge large amounts of vapor from the porous medium, particularly under high heat flux conditions of 10 MW/m^2 level. If the upstream pressure of high-speed vapor flow exceeds a pumping pressure/ an inlet pressure shortly before the porous medium, the vapor phase near the heat transfer surface particularly starts to excessively grow and form a thermal resistance layer that quickly deteriorates the heat transfer performance. In fact, in the heat transfer tests by the author in which the sintered-metal-particles and the sintered-metal-fibers were used, such phenomena have been confirmed that the heat transfer performance of the porous media with higher permeability is reversed under high heat flux conditions exceeding several MW/m^2 [31, 40]. It is effective to yield a multiple structure to the porous medium (e.g., biporous structure at the lower right of **Figure 1**) to provide a path for vapor discharge, however, it is extremely important to reduce the flow resistance in the porous medium because the vapor is first generated within the porous medium. In fact, the use of microchannels is also important in this technology as we want to promote evaporative heat transfer [41]. Therefore, effecting compromise with curved flow paths such as in the sintered-particles and introducing unidirectional porous media provide powerful means to improve the permeability. Here, to evaluate the permeability of the unidirectional porous media, the absolute permeability is modeled based on the Darcy–Weithbach equation, which gives the pressure loss of a circular pipe flow (the inflow resistance to the porous medium and the outflow resistance are ignored).

$$K = \varepsilon d_p^2 / 32 \tag{4}$$

Figure 7 shows the permeability of the sintered-particles and the unidirectional porous media. The horizontal axis is the porosity with a characteristic length d_p of 100 μm (corresponding to the particle diameter for the sintered-particles and the pore diameter for the unidirectional porous media). Indeed, with increase in porosity, the permeability increases; however, for example, if the porosity is 0.3, 0.4, and 0.5, the permeability ratios of the unidirectional porous media and the sintered-particles are 30.6, 12.7, and 5.6 times, respectively. This ratio does not depend on the characteristic

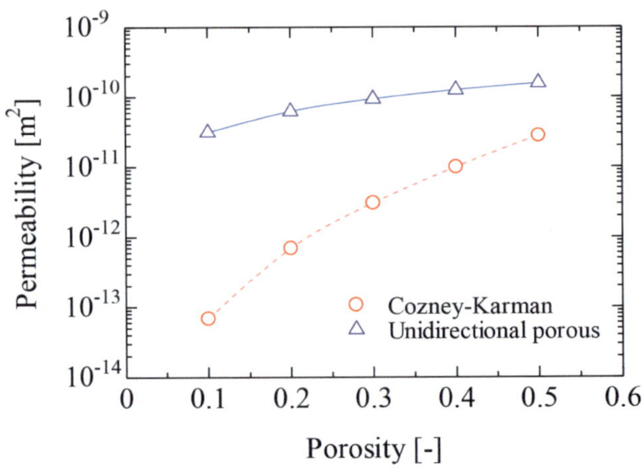

Figure 7.
Permeability of unidirectional porous.

length. In other words, when focusing on the porous structure, the unidirectional porous media can be considered as the one that offers the best compromise between the improvement of the thermal conductivity and the improvement of the permeability. An example of application is a fibrous wick manufactured by sintering fibers, which considerably reduces the flow resistance in the fiber direction to contribute to the improvement in the maximum heat transport capacity of small heat pipes [3].

2.3 Other heat transfer-related issues for introducing porous media

An important issue when introducing the porous media as a heat transfer promoter is the contact thermal resistance generated at the contact interface between the porous medium and the heat transfer surface. In general, with increasing heat flux, the temperature gap increases due to the increase in contact thermal resistance generated at the interface, so that the heat transfer coefficient is directly affected. For instance, when a sintered-particle is mechanically pressed to the heat transfer surface, a large contact thermal resistance is generated because of the point contact state between the heat transfer surface and the particles. In the point contact state, the heat conduction to the first particle is considerably deteriorated, so that boiling and evaporation occur mainly on the heat transfer surface or the first several layers of the particle, and thus, effective utilization of latent heat of vaporization inside the porous medium cannot be expected. The best measure is sintering the sintered-particle to the heat transfer surface. Remarkable research results are reported by Kibushi et al. [42] regarding the contact thermal resistance in a high heat flux environment exceeding 1 MW/m^2 (for the experimental details refer to the reference). **Figure 8** shows the temperature gap that occurs at the contact interface of two flat surfaces. The loads on a jointing surface of ϕ10 mm are 0.33, 1.71, and 3.03 MPa, and there are the two patterns of mechanical and solder joints. Focusing on the heat flux of 2 MW/m^2, in the mechanical joint case, a temperature difference of ~40 K occurs under a load of 3.03 MPa, and a temperature gap exceeding 100 K takes place at the load of 0.33 MPa. However, it can also be confirmed that the contact thermal resistance is significantly improved by solder bonding. At the heat flux of 5 MW/m^2, the temperature gap is ~10 K, and it is ~20 K upon linear extrapolation to 10 MW/m^2. Accordingly, it is evident that when the porous medium is loaded on the heat transfer surface, the contact thermal resistance should be reduced by solder bonding or the like. Our heat transfer tests regarding a gas impinging jet flow into a sintered-particle verified that the heat transfer coefficient is spectacularly increased by soldering the

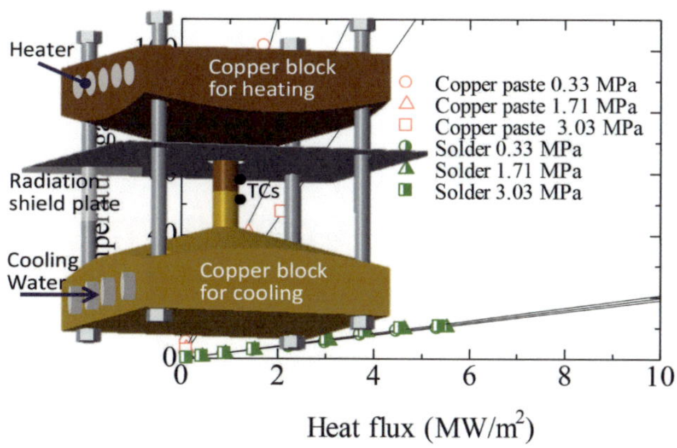

Figure 8.
Temperature gap due to contact thermal resistance.

porous medium to the heat transfer surface [43]. Further, the pool boiling results by Peterson et al. [44] have also demonstrated that boiling heat transfer is dramatically enhanced by HIP bonding of the porous medium to the heat transfer surface.

For introducing the unidirectional porous media proposed this time, if the pores that serve as the flow path are parallel to the heat transfer surface, a perfect surface contact state can be achieved over the entire heat transfer surface. Furthermore, even if the pores are perpendicular to the heat transfer surface, the surface contact state can be maintained in the solid phase portion, which can maximally reduce the contact thermal resistance between the porous medium and the heat transfer surface.

2.4 Introduction of unidirectional porous copper

A majority of unidirectional porous media proposed in this study is generally fabricated by electric discharge machining or MEMS. As the production with these techniques is very expensive, it is desirable to introduce the unidirectional porous media with excellent mass productivity. Here, three unidirectional porous media are presented, which are currently introduced to the author's laboratory. The first is the lotus copper shown in **Figure 9(a)**, in which pores are formed utilizing different saturation degree of dissolved hydrogen existing in molten copper with temperature gradient [45]. Chiba et al. already demonstrated its superiority in single-phase flow heat transfer [46]. The boiling heat transfer enhancement technology that the author is currently developing in cooperation with Lotus Thermal Solutions Co., Ltd. will be introduced in the next section. Hokamoto et al. also proposed a technique for forming a group of thin metal tubes into a unidirectional porous tube by exploded welding technique (**Figure 9(b)** [47]). To date, the author has introduced unidirectional porous copper tubes as a heat transfer promoter of a gas flow in a joint research with Hokamoto et al. of Kumamoto University. Based on single-phase flow heat transfer tests, the unidirectional porous tubes of 21 mm in outer diameter

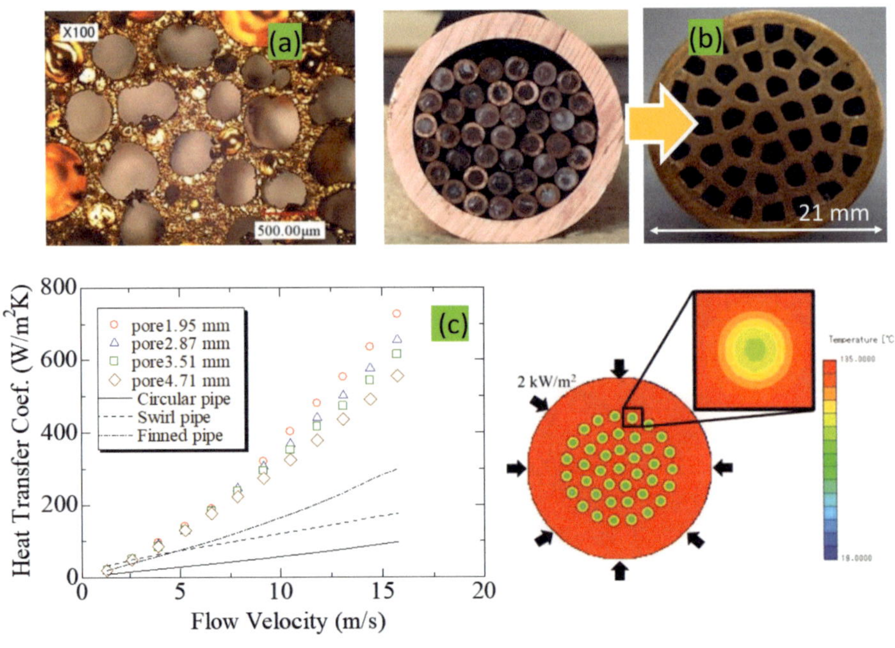

Figure 9.
Unidirectional porous media ((a) lotus copper, (b) exploded welded porous pipe) and (c) heat transfer performance of a gas flow in exploded welded porous pipes & its CFD simulation.

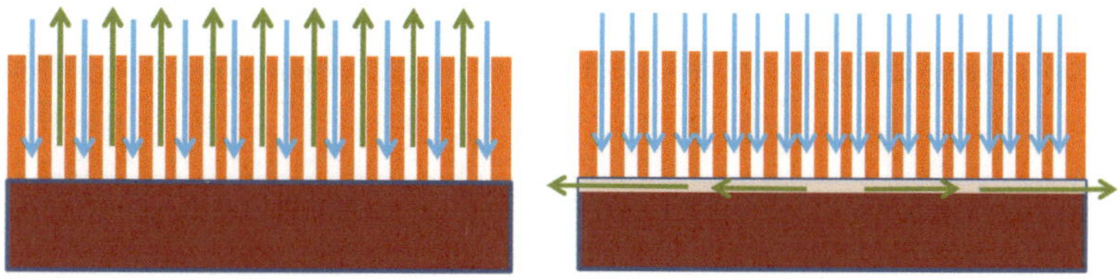

Figure 10.
Two heat removal devices proposed.

shows a heat transfer performance that is up to approximately eight times higher than that of a smooth tube at the maximum as shown in **Figure 9(c)** [48]. The pore size and the porosity of this porous tube can be adjusted by varying the inner diameter and wall thickness of the thin tube. The lastly introduced unidirectional porous copper is the one molded by a metal 3D printing technique [49]. Recent advances in 3D printing technology enabled the molding of even porous media composed of copper and having microchannels. The adjustment of the pore diameter and the pore structure matching the heat transfer mechanism are possible, which is the greatest attraction of this technology.

On the other hand, when this type of unidirectional porous copper is applied as a promoter of boiling/evaporation heat transfer, particularly if the pores are perpendicular to the heat transfer surface, one of the advantages is that the coolant can surely be supplied to the heat transfer surface. The major limitation is how to rapidly discharge a large amount of vapor generated in the porous medium to the outside. As a solution, we propose to joint the unidirectional porous copper to the grooved heat transfer surface; thus, as shown in **Figure 10**, two cooling structures are enabled. The first structure on the left shows the liquid supply direction is opposite to the vapor discharge direction, whereas the second structure on the right shows a method of directly discharging the vapor outside the porous medium via these grooves. In the next section, we will focus on the cooling performance related to flow boiling heat transfer and pool boiling heat transfer using these cooling structures.

3. Enhancement of flow boiling and pool boiling heat transfers using unidirectional porous copper

3.1 Enhancement of flow boiling/evaporative heat transfer with EVAPORON-4

As mentioned in the Introduction, the author proposed EVAPORON, EVAPORON-2, and EVAPORON-3, cooling devices using porous metals [29–34]. The porous metal is connected to the cooling surface as shown in **Figure 11**, the cooling liquid is supplied into the porous medium in a countercurrent to the heat flow, and the heat is removed by the vigorous phase change of the cooling liquid within. To smoothly discharge the large amount of vapor outside the porous medium, EVAPORON-2 has several subchannels inside the porous medium, and EVAPORON-3 has a liquid–vapor-separator on the porous medium. Here, the concept of cooling device EVAPORON-4, using a unidirectional porous media is realized as shown in **Figure 12(a)**. As shown in **Figure 12(b)**, 9 × 5 grooves for the vapor discharge are formed on the heat transfer surface, and the unidirectional porous copper shown in **Figure 12(c)** is joined by soldering. The groove is 1.0 mm in width and 0.5 mm in depth. The unidirectional porous copper with a diameter of 20 mm and a thickness of 10 mm is produced by machining and has 248 small pores of ϕ0.5 mm in diameter. In addition, the porous medium has large five ϕ2.6 mm holes for the vapor discharge. EVAPORON-4 is a once-through type cooling device in which the cooling liquid supplied from the upper part of the unidirectional porous copper through the small pores undergoes phase change in the pores and the grooves to be then discharged through the grooves and the vapor discharging large holes. **Figure 13** shows some of the results in two cases that the unidirectional

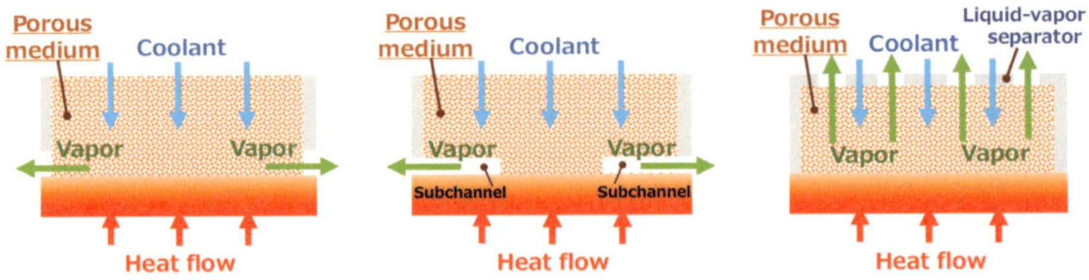

Figure 11.
From the left: EVAPORON, EVAPORON-2, and EVAPORON-3.

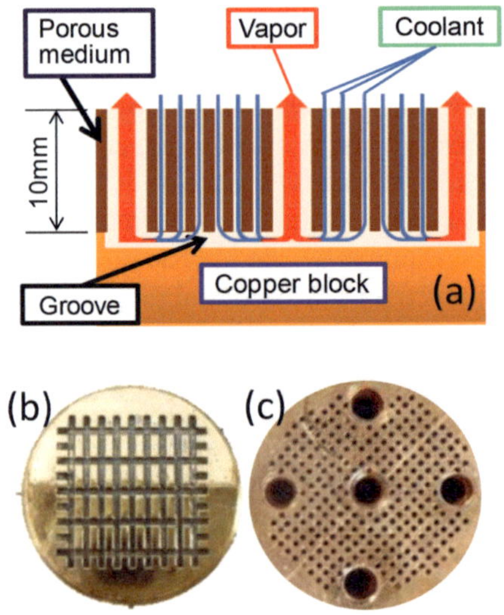

Figure 12.
Outline of EVAPORON-4.

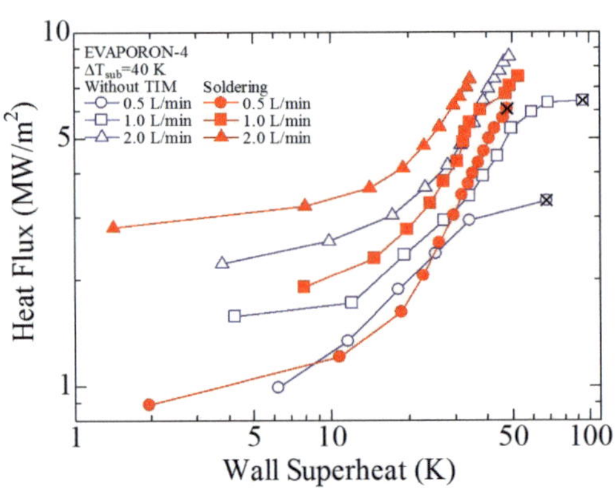

Figure 13.
Boiling curves of EVAPORON-4.

porous copper is jointed onto the heat transfer surface with/without soldering ("Without TIM (Thermal Interface Material)" shows the data of mechanical joint without bonding). For details of the experimental apparatus and various results, please refer to Reference [50]. The inlet liquid subcooling of distilled water is 40 K, whereas the x mark on the plot shows that the obtained data represent the critical heat flux. From this figure, the data with soldering move to the lower wall-superheat side under each flow rate condition, which verifies that the phase change heat transfer is enhanced by solder bonding the porous copper to the grooved heat transfer surface. Focusing on the data obtained at the flow rate of 0.5 L/min, while the critical heat flux without solder bonding is ~3 MW/m^2, the critical heat flux is improved to exceed 6 MW/m^2 by solder bonding (1.8-fold enhancement). For the flow rate of 2.0 L/min, as the maximum heat flux (not the critical heat flux) without solder bonding is approximately 9 MW/m^2, the critical heat flux by solder bonding is expected to exceed 10 MW/m^2, though the experiment was stopped due to the temperature limit of the cartridge heaters we used. For reference, the measured critical heat flux of the impinging jet at the flow rate of 2.0 L/min is approximately 4 MW/m^2 (the same flow rate is ejected from a hole of ϕ2 mm), so that the cooling performance of EVAPORON-4 is much higher compared with other cooling technologies. These results are attributed to the effective functioning of the phase change within and the vapor discharge outside the porous medium, because the heat transfer is remarkably improved under high heat flux conditions.

Currently, we are examining the flow rate distribution and the phase change in porous media by CFD and visualization experiments. From now onward, we intend to push forward discussions and optimization by introducing the porous copper fabricated by a 3D metal printer, as shown in **Figure 14**, thus comprising the un-uniform unidirectional pore and groove structures that can make maximum use of the latent heat of vaporization [49]. Moreover, we are examining a small-sized cold plate having in mind an application to electronic devices, a theme to be referred to in Reference [51].

3.2 Enhancement of critical heat flux of saturated pool boiling using lotus coppers by "breathing phenomenon"

In the past heat transfer experiments using EVAPORONs, the author noticed that the inlet pressure of the fluid rapidly decreases when the phase change in the porous medium is considerable and the vapor is vigorously ejected to the outside, i.e., there must be spontaneous liquid supply phenomenon associated with the discharge of vapor (henceforth referred to as "Breathing Phenomenon"). As shown in **Figure 15**, the author has proposed to attach a porous plate called "lotus copper" with a unidirectional pore structure to the grooved heat transfer surface (**Figure 9(a)**), whereby the critical heat flux of saturated pool boiling is spectacularly increased [15, 52–54].

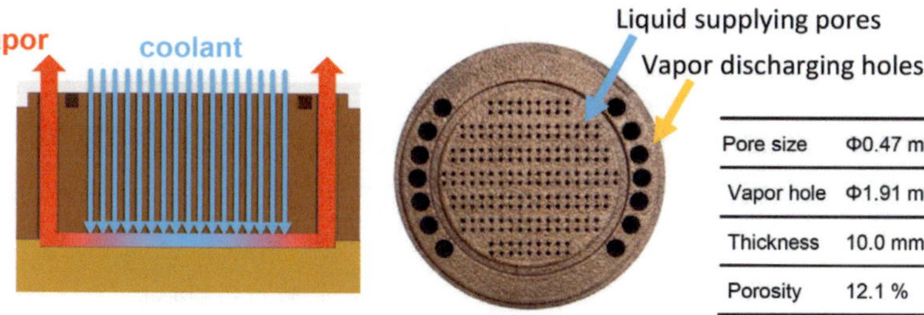

Pore size	Φ0.47 mm×202
Vapor hole	Φ1.91 mm×14
Thickness	10.0 mm
Porosity	12.1 %

Figure 14.
Unidirectional porous copper fabricated by 3D printing technique.

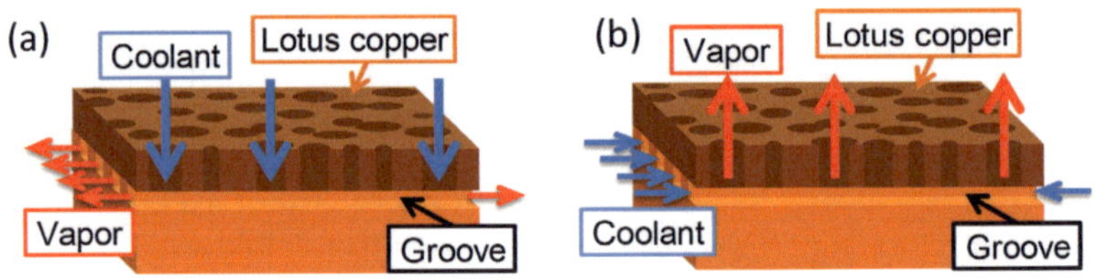

Figure 15.
Breathing phenomenon spontaneously induced by lotus copper on a grooved heat transfer surface ((a) mode-G, (b) mode-L).

This two-phase immersion cooling technique we usually call "Lotus' Breathing" offers the following features.

1. Spontaneous liquid supply phenomenon due to vapor blowout (Breathing Phenomenon)

2. No requirement of capillarity induced in the porous medium

3. Once-through type liquid supply and vapor discharge (**Figure 15**). Liquid supply and vapor discharge are separated and formation of coalesced bubbles is retarded.

4. Use of a unidirectional porous copper with high permeability and high thermal conductivity to enhance the breathing phenomenon and the heat transfer in the porous medium.

There are two possible breathing modes "Mode-G" and "Mode-L", as in **Figure 15(a)** and **(b)**, i.e., the Mode-G on the left in which the vapor is discharged from the groove and accordingly the liquid is supplied from the upper part of the porous medium as well as the inverse mode-L.

To demonstrate the increase in CHF by the breathing phenomenon, saturated pool boiling experiments have been conducted in an atmospheric pressure environment. For details of the experiment, please refer to Reference [15]. The Lotus copper plate (10 mm x 10 mm, thickness = 2 mm, porosity = 65.9%, and average pore diameter = 0.49 mm) is attached to a 10 mm × 10 mm heat transfer surface provided with 0.5 mm-square or 1.0 mm-square unidirectional grooves as a boiling heat transfer surface. The results for water and FC72 are shown in **Figure 16**. In the case of water, the critical heat flux for the smooth surface in this device is 1.4 MW/m^2,

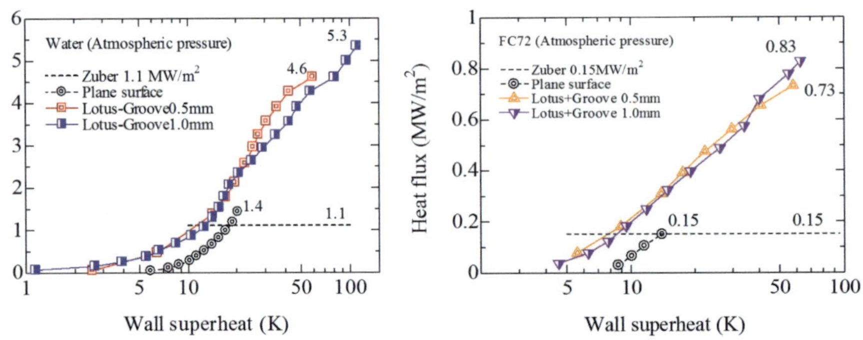

Figure 16.
CHF improvement by breathing phenomenon (left: Water, right: FC72).

which is an adequate value for a boiling heat transfer experiment using the heat transfer block. The results also verify that the critical heat flux when utilizing the lotus copper jointed on the grooved heat transfer surface considerably improves by providing wider grooves. The 0.5 mm-square groove achieves the critical heat flux of 4.6 MW/m^2 at the wall superheat of 58.9 K, and the one of 1.0 mm-square groove achieves the critical heat flux of 5.3 MW/m^2 at the wall superheat of 111 K, which indicates that the groove size is an important factor for improving the critical heat flux by the breathing phenomenon. The boiling heat transfer is also remarkably enhanced even if compared to that for the smooth surface; this technology has proven to achieve both the enhancement of the boiling heat transfer and the critical heat flux. By visualizing the boiling phenomenon shortly before the critical heat

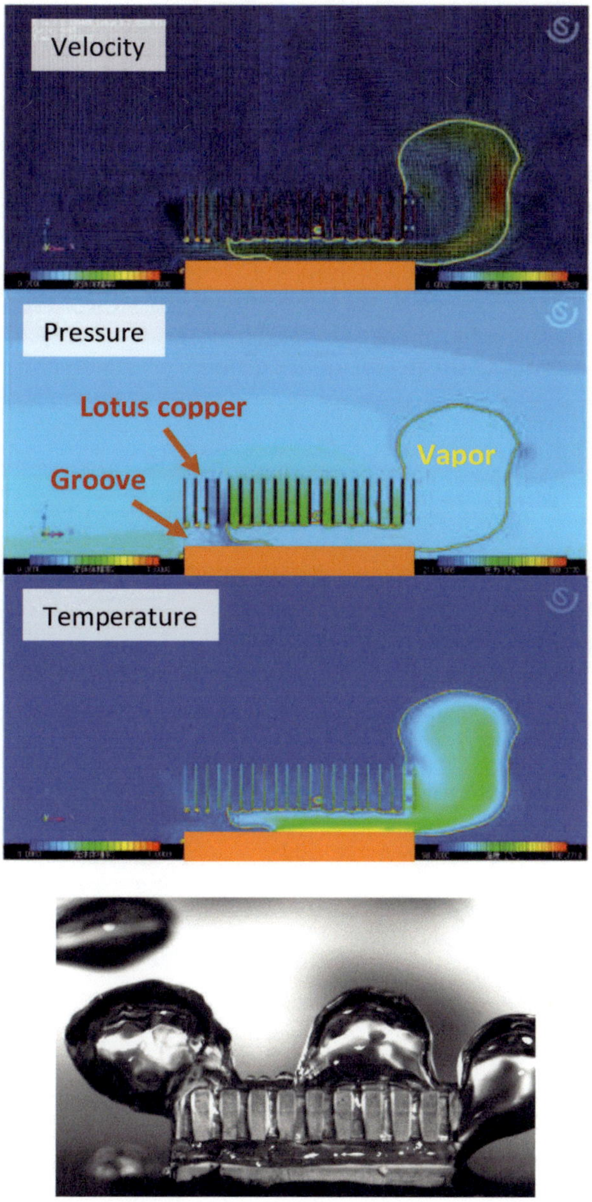

Figure 17.
Example of CFD simulation and visualization experiment of breathing phenomenon.

flux, a large amount of vapor is discharged from the grooves. Thus, the breathing phenomenon in these experiments conceivably corresponds to the Mode-G of **Figure 15(a)**. On the other hand, in the FC72 case, the maximum heat flux, which is not the critical heat flux, is 0.83 (MW/m^2) that is 5.5 times higher than CHF of the smooth surface. This result verifies that the breathing phenomenon can work for dielectric fluid and increases the critical heat flux.

Currently, the mechanism of critical heat flux enhancing and the factors that further entails the breathing phenomenon are being discussed based on the visualization experiment and CFD simulation as shown in **Figure 17**. In addition, the effects of varying the porosity, the pore size of the lotus copper, and the groove structure have been evaluated [55, 56].

4. Conclusion

In this chapter, the thermal conductivity and permeability of various porous media were quantitatively evaluated; as one of the porous media appropriate for high heat flux heat removal, targeted by the author, the introduction of "porous metal with unidirectional pores" was strongly suggested. Furthermore, as an application of using unidirectional porous coppers, we demonstrated the cooling performance of the flow boiling/evaporation cooling device EVAPORON-4, and that of the pool boiling enhancing technology "Lotus Cooler" based on "Breathing Phenomenon". From now onward, we intend to optimize the structure of the unidirectional porous media and the grooved heat transfer surface, thus aiming at improving and controlling the performance of these cooling technologies.

Author details

Kazuhisa Yuki
Tokyo University of Science, Yamaguchi, Japan

*Address all correspondence to: kyuki@rs.socu.ac.jp

References

[1] Japan association for heat pipe, Practical heat pipe 2nd edition, Nikkan Kogyo Shimbun (2001).

[2] Gaurav Patankar, Justin A. Weibel, Suresh V. Garimella, On the transient thermal response of thin vapor chamber heat spreaders: Optimized design and fluid selection, International Journal of Heat and Mass Transfer, 148, 119106 (2020).

[3] Fujikura Ltd., https://www.fujikura.co.jp/products/electronics/thermal/01/2044118_12671.html

[4] Kazuhisa Yuki, Boiling -Research and Advances-, Elsevier, pp.593-609 (2017).

[5] Shoji Mori, Yoshio Utaka, Critical heat flux enhancement by surface modification in a saturated pool boiling: A review, International Journal of Heat and Mass Transfer, vol. 108, pp. 2534-2557 (2017).

[6] Shoji Mori, Boiling -Research and Advances-, Elsevier, pp.212-226 (2017).

[7] Chinmay M. Patil, Satish G. Kandlikar, Review of the manufacturing techniques for porous surfaces used in enhanced pool boiling, Heat Transfer Engineering, 35, No. 10, pp.887-902 (2014).

[8] Shoji Mori, Kunito Okuyama, Enhancement of the critical heat flux in saturated pool boiling using honeycomb porous media, International Journal of Multiphase Flow, 35, 10, pp. 946-951 (2009).

[9] Md Mahamudur Rahman, Emre Ölçeroğlu, and Matthew McCarthy, Role of wickability on the critical heat flux of structured superhydrophilic surfaces, Langmuir, 30, 37, pp. 11225-11234 (2014).

[10] Calvin H. Li, T. Li, Paul Hodgins, Chad N. Hunter, Andrey A. Voevodin, John G. Jones, G.P. Peterson, Comparison study of liquid replenishing impacts on critical heat flux and heat transfer coefficient of nucleate pool boiling on multiscale modulated porous structures, International Journal of Heat and Mass Transfer, 54, 15-16, pp. 3146-3155 (2011).

[11] Scott G Liter, Massoud Kaviany, Pool-boiling CHF enhancement by modulated porous-layer coating: theory and experiment, International Journal of Heat and Mass Transfer, 44, 22, pp. 4287-4311 (2001).

[12] Xianming Dai, Xinyu Huang, Fanghao Yang, Xiaodong Li, Joshua Sightler, Yingchao Yang, and Chen Li, Enhanced nucleate boiling on horizontal hydrophobic-hydrophilic carbon nanotube coatings, Applied Physics Letters 102 (16), 161605 (2013).

[13] Xianming Dai, Pengtao Wang, Fanghao Yang, Xiaochuan Li, and Chen Li, Decoupling the influence of surface structure and intrinsic wettability on boiling heat transfer, Applied Physics Letters, 112, 253901 (2018).

[14] Lizhan Bai, Lianpei Zhang, Guiping Lin, and G. P. Peterson, Pool boiling with high heat flux enabled by a porous artery structure, Applied Physics Letters, 108, 233901 (2016).

[15] Kazuhisa Yuki, Risako Kibushi, Rikako Tsuji, Kio Takai, Noriyuki Unno, Tetsuro Ogushi, Masaaki Murakami, Tomiyuki Numata, Hikaru Nomura, Takuya Ide, Thermal management of auto motive SiC-based on-board inverter with 500 W/cm^2 in heat flux, and Two-phase immersion cooling by breathing phenomenon spontaneously induced by lotus porous copper jointed onto a grooved heat transfer surface, Journal of Thermal Science and Technology, Vol.15, No.1, Paper No.20-00051 (2020).

[16] Shadi Mahjoob, KambizVafai, A synthesis of fluid and thermal transport

models for metal foam heat exchangers, International Journal of Heat and Mass Transfer, 51, 15-16, pp. 3701-3711 (2008).

[17] Z.F. Huang, A. Nakayama, K. Yang, C. Yang, W. Liu, Enhancing heat transfer in the core flow by using porous medium insert in a tube, International Journal of Heat and Mass Transfer, 53, 5-6, pp. 1164-1174 (2010).

[18] Kouichi Kamiuto, San San Yee, Heat transfer correlations for open-cellular porous materials, International Communications in Heat and Mass Transfer, 32, 7, pp. 947-953 (2005).

[19] T. Takeda, D. Sampei, R. Miura, Heat transfer and fluid flow characteristics of a heat exchanger with highly porous thin copper wire inserted, Proceedings of the 30th International Symposium on Transport Phenomena (ISTP30), ISTP080 (2019)

[20] R. M. Fand, M. Varahasamy, L.S. Greer, Empirical correlation equations for heat transfer by forced convection from cylinders embedded in porous media that account for the wall effect and dispersion, International Journal of Heat and Mass Transfer, 36, 18, pp. 4407-4418 (1993).

[21] Kazuhisa Yuki, Masumi Okumura, Hidetoshi Hashizume, Saburo Toda, Neil B. Morley, Akio Sagara, Flow visualization and heat transfer characteristics for sphere-packed pipes, Journal of Thermophysics and Heat Transfer, 22, 4, pp. 632-648 (2008).

[22] S. Sharafat, A. Mills, D. Youchison, R. Nygren, B. Williams & N. Ghoniem, Ultra low pressure-drop helium-cooled porous-tungsten PFC, Journal Fusion Science and Technology, 52, 3, pp. 559-565 (2007).

[23] S. Sharafat, N. Ghoniem, M. Sawan, A. Ying, B. Williams, Breeder foam: an innovative low porosity solid breeder material, Fusion Engineering and Design, 81(1), pp. 455-460 (2006).

[24] Mohamed H. Nasr, Craig E. Green, Peter A. Kottke, Xuchen Zhang, Thomas E. Sarvey, Yogendra K. Joshi, Muhannad S. Bakir, Andrei G. Fedorov, Hotspot thermal management with flow boiling of refrigerant in ultrasmall microgaps, Journal of Electronic Packaging, 139/011006 (2017).

[25] Mohamed H. Nasr, Craig E. Green, Peter A.Kottke, Xuchen Zhang, Thomas E. Sarvey, Yogendra K. Joshi, Muhannad S. Bakir, Andrei G. Fedorov, Flow regimes and convective heat transfer of refrigerant flow boiling in ultra-small clearance microgaps, International Journal of Heat and Mass Transfer, 108, pp. 1702-1713 (2017).

[26] Pouya Asrar, Xuchen Zhang, Craig E. Green, Muhannad Bakir, Yogendra K. Joshi, Flow boiling of R245fa in a microgap with staggered circular cylindrical pin fins, International Journal of Heat and Mass Transfer, 121, pp. 329-342 (2018).

[27] Shankar Narayanan, Andrei G. Fedorov, Yogendra K. Joshi, Heat and mass transfer during evaporation of thin liquid films confined by nanoporous membranes subjected to air jet impingement, International Journal of Heat and Mass Transfer, 58, pp. 300-311 (2013).

[28] Przemysław Smakulski, Sławomir Pietrowicz, A review of the capabilities of high heat flux removal by porous materials, microchannels and spray cooling techniques, Applied Thermal Engineering, 104, pp. 636-646 (2016).

[29] S. Toda, K. Yuki, S. Ebara, Y. Kunikata, *J. abei*, H. Hashizume, Development of an advanced cooling device using porous media with active boiling flow counter to high heat flux, Proceedings of the 13th International

Heat Transfer Conferences (IHTC13), BOI-58 (2006).

[30] Kazuhisa Yuki, Jun Abei, Hidetoshi Hashizume and Saburo Toda, Super-high heat flux removal using sintered metal porous media, Journal of Thermal Science, 14, 3, pp. 272-280 (2005).

[31] K. Yuki, K. Suzuki , H. Hashizume, S. Toda, Key issues to enable heat flux removal exceeding $10MW/m^2$ by use of metal porous media as latent-heat transfer device, Special Topics & Reviews in Porous Media — An International Journal, vol. 1, no. 1, pp. 1-13 (2010).

[32] Kazuhisa Yuki, Hidetoshi Hashizume, Saburo Toda, Sub-channels-inserted porous evaporator for efficient divertor cooling, Fusion Science and Technology, 60, pp. 238-242 (2011).

[33] Kazuhisa Yuki, Hidetoshi Hashizume, Saburo Toda, Akio Sagara, Divertor cooling with sub-channels-inserted metal porous media (Effects of the sub-channel volume and location on heat transfer characteristics), Fusion Science and Technology, 64, 2, pp. 325-330 (2013).

[34] Kio Takai, Kazuhisa Yuki, Akio Sagara, Heat transfer performance of EVAPORON-3 developed for an enlarged heat transfer surface of Divertor, Plasma and Fusion Research, 12, 1405015 (2017).

[35] Kazuhisa Yuki, Kio Takai, Yoshiki Indo and Koichi Suzuki, Proposal of utilizing uni-directional porous metal for extremely high heat flux removal, Proceedings of the 6th International Conference on Porous Media and Its Applications in Science and Engineering (ICPM6), (2016).

[36] Noriyuki Unno, Kazuhisa Yuki, Ryo Inoue, Yasuo Kogo, Jun Taniguchi, Shin-ichi Satake, Enhanced evaporation of porous materials with micropores and high porosity, Journal of Thermal Science and Technology, 15, 1, Paper No.19-00654 (2020).

[37] K. Boomsma, D. Poulikakos, On the effective thermal conductivity of a three-dimensionally structured fluid-saturated metal foam, International Journal of Heat and Mass Transfer, 44, 4, pp. 827-836 (2001).

[38] T. Ogushi and H. Chiba, H. Nakajima and T. Ikeda, Measurement and analysis of effective thermal conductivities of lotus-type porous copper, Journal of Applied Physics, 95, 10, pp. 5843-5847 (2004).

[39] Hiroshi Chiba, Tetsuro Ogushi, Hideo Nakajima, Teruyuki Ikeda, Heat transfer capacity of lotus-type porous copper heat sink, JSME International Journal, Series B Fluids and Thermal Engineering, 47, 3, pp. 516-521 (2004).

[40] Kazuhisa Yuki, Performance improvement and heat countermeasure technology for high temperature resistant and high heat dissipation members, Technical Information association, (2017) (in Japanese).

[41] Shu Soma, Tomoaki Kunugi, Phenomenological model for non-isothermal capillary evaporation in narrow channel, International Journal of Multiphase Flow, 122, 103154 (2020).

[42] R. Kibushi, K. Yuki, N. Unno, K. Yuki, T. Tomimura, T. Hatakeyama, M. Ishizuka, Thermal resistance evaluation in high heat flux electronics, Proceedings of 2018 International Conference on Electronics Packaging and iMAPS All Asia Conference (ICEP-IAAC2018), FE2-3 (2018).

[43] K. Yuki, M. Kawamoto, S. Nagano, K. Yuki, K. Suzuki, 53rd Japan Heat Transfer Symposium, G113 (2016) (in Japanese).

[44] Choongho Yu, Li Shi, Alan M. Cassell, Brett A. Cruden, Quoc Ngo,

Jun Li, Thermal contact resistance and thermal conductivity of a carbon nanofiber, Proceedings of 2005 ASME Summer Heat Transfer Conference (2005).

[45] Hideo Nakajima, Fabrication, properties and application of porous metals with directional pores, Progress in Materials Science, 52, 7, pp. 1091-1173 (2007).

[46] H. Chiba, T. Ogushi, H. Nakajima, Heat transfer capacity of lotus-type porous copper heat sink for air cooling, Journal of Thermal Science and Technology (Journal of Thermal Science and Technology, 5, 2, pp. 222-237 (2010).

[47] K. Hokamoto, M. Vesenjak, Z. Ren, Fabrication of cylindrical uni-directional porous metal with explosive compaction, Materials Letters, 137, pp. 323-327 (2014).

[48] K. Yuki, Y. Sato, R. Kibushi, N. Unno, K. Suzuki, T. Tomimura and K. Hokamoto, Heat transfer performance of porous copper pipe with uniformly-distributed holes fabricated by explosive welding technique, Proceedings of the 27th International Symposium on Transport Phenomena, Paper no. 126 (2016).

[49] K. Takai, D. Suga, K. Yuki, R. Kibushi, N. Unno, K. Shimamoto, Development of Cooling Device utilizing, Uni-directional Porous Copper Manufactured by Metal 3D Printer Proceedings of the 56th Japan Heat Transfer Symposium, I114 (2019) (in Japanese).

[50] Kio Takai, Kohei Yuki, Kazuhisa Yuki, Risako Kibushi, Noriyuki Unno, Teruya Tanaka, Heat transfer performance of an energy-saving heat removal device with uni-directional porous copper for divertor cooling, Fusion Engineering and Design, 136, Part A, pp. 518-521 (2018).

[51] K. Yuki, R. Tsuji, K. Takai, R. Kibushi, N. Unno, K. Suzuki, Development of a small cooling device with uni-directional porous copper, Proceedings of the 54th Japan Heat Transfer Symposium, C115 (2017) (in Japanese).

[52] Kazuhisa Yuki, Tomohiro Hara, Soichiro Ikezawa, Kentaro Anju, Koichi Suzuki, Tetsuro Ogushi, Takuya Ide, Masaaki Murakami, Immersion Cooling of Electronics utilizing Lotus-Type Porous Copper, Transactions of The Japan Institute of Electronics Packaging, 9 (2016), pp. E16-013-1-E16-013-7.

[53] K. Yuki, K. Takai, K. Anju, R. Kibushi, N. Unno, T. Ogushi, M. Murakami, T. Ide, Thermal management of electronics by uni-directional porous heat sinks (CHF enhancement of saturated pool boiling toward immersion cooling of on-vehicle inverter), 2017 Annual Meeting of the Japan Society of Mechanical Engineers, J0330103 (2017) (in Japanese).

[54] Yuki, K., Kibushi, R., Unno, N., Ide, T., Ogushi, T., Murakami, M., Patent publication number 2018-204882 (2018).

[55] D. Negita, K. Yuki, N. Unno, R. Kibushi, T. Ide, T. Ogushi, M. Murakami, T. Numata, H. Nomura, Critical heat flux improvement of two-phase immersion cooling by controlling breathing phenomenon, Proceedings of the International Technical Conference and Exhibition on Packaging and Integration of Electronic and Photonic Microsystems (InterPACK2020-Online), (2020).

[56] Daiki Suga, Kazuhisa Yuki, Risako Kibushi and Kio Takai, Visualization of two-phase flow in uni-directional porous copper, Proceedings of the the 31st International Symposium on Transport Phenomena (ISTP31), Paper no. 128, (2020).

Section 5

Modelling of Porous Media and Systems

Chapter 6

Inverse Measurement of the Thickness and Flow Resistivity of Porous Materials via Reflected Low Frequency Waves-Frequency Approach

Mustapha Sadouki

Abstract

A direct and inverse method is proposed for measuring the thickness and flow resistivity of a rigid air-saturated porous material using acoustic reflected waves at low frequency. The equivalent fluid model is considered. The interactions between the structure and the fluid are taken by the dynamic tortuosity of the medium introduced by Johnson et al. and the dynamic compressibility of the air introduced by Allard. A simplified expression of the reflection coefficient is obtained at very low frequencies domain (Darcy's regime). This expression depends only on the thickness and flow resistivity of the porous medium. The simulated reflected signal of the direct problem is obtained by the product of the experimental incident signal and the theoretical reflection coefficient. The inverse problem is solved numerically by minimizing between simulated and experimental reflected signals. The tests are carried out using two samples of polyurethane plastic foam with different thicknesses and resistivity. The inverted values of thickness and flow resistivity are compared with those obtained by conventional methods giving good results.

Keywords: acoustic characterization, porous materials, fluid equivalent model, reflected wave, Darcy's regime

1. Introduction

Porous materials are of great importance for a wide range of industrial and engineering applications, including transportation, construction, aerospace, biomedical and others. These materials, such as plastic foams, fibers and granular materials are frequently used for sound and heat insulation in buildings, schools and hospitals to minimize noise and reduce nuisance.

The propagation of sound in a porous material is a phenomenon that governed by physical characteristics of a porous medium. Porous sound absorbers are materials in which sound propagation takes place in a network of interconnected pores such that the viscous and thermal interaction causes the dissipation of acoustic energy and converts it into heat. Knowledge of the acoustic and physical properties of these materials is of great importance in predicting their acoustic behavior and their insulate ability against noise and heat. For this reason, there are many works of research and studies in the literature [1–15] that are articulated in this line of inquiry where many mathematical and semi-phenomenological models have been developed to study the acoustic behavior of these materials. Among the most important of these models, we find the JCA model (Johnson-Champoux-Allard model) [1–4] used in the case of porous materials with a rigid structure saturated with air.

According to the JCA model [3, 4], The acoustic propagation in air saturated porous materials is described by the inertial, viscous, and thermal interactions between the fluid and the structure [1–5]. In the high frequency domain [1–4] the inertial, viscous and thermal interactions are taken into account, by the high limit of tortuosity for the inertial effects [3], and by the viscous and thermal characteristic length [1, 2, 4] for the viscous and thermal effects. In the low-frequency domain [1, 2, 11, 13], inertial, viscous and thermal interactions are described by the inertial and thermal tortuosity and by the viscous and thermal permeability. In very low frequency approximation, the viscous-inertial interactions [11, 14, 15] are only described by the flow resistivity. The determination of these parameters is crucial for the prediction of sound damping in these materials.

The objective of this work is to propose an acoustic method based on the resolution of the direct and inverse problem using reflected acoustic waves at low frequency to determine the thickness and flow resistivity describing the porous medium. The direct problem consists in constructing theoretically the reflected signal knowing the incident signal and the parameters of the medium; given the experimental incident signal denoted by $\boldsymbol{p^i_{exp}}(\boldsymbol{\omega})$, and the reflection coefficient which plays the role of a transfer function of the medium denoted $\boldsymbol{R}(\boldsymbol{\sigma}, \boldsymbol{L}, \boldsymbol{\omega})$ as a function of the parameters to be found, we deduce the simulated reflected signal $\boldsymbol{p^r_{sim}}(\boldsymbol{\sigma}, \boldsymbol{L}, \boldsymbol{\omega})$ which must be compared to the experimental reflected signal $\boldsymbol{p^r_{exp}}(\boldsymbol{\omega})$. The inverse problem therefore consists in minimizing the difference between the $\boldsymbol{p^r_{exp}}(\boldsymbol{\omega})$ and $\boldsymbol{p^r_{sim}}(\boldsymbol{\sigma}, \boldsymbol{L}, \boldsymbol{\omega})$ signals by varying the required parameters. The solution corresponds to the sets of parameters that give the minimum deviation between the simulated reflected signal and the experimental reflected signal.

2. Acoustical model

The porous material is a bi-phasic medium consisting of a solid part and a fluid part that saturates the pores. When the solid part is flexible, the two phases start moving simultaneously under excitation by an acoustic wave; in this case the dynamics of the movement is well described by Biot's theory [16–18]. In the case of a rigid material, the solid part remains immobile and the acoustic waves propagate only in the fluid. This case is described by the equivalent fluid theory [1–5]. In this theory the viscous and inertial interactions within the medium are described by the dynamic tortuosity introduced by Johnson et al. [2, 3] while the thermal effects are taken into account by the dynamic compressibility of the fluid given by Allard and Champoux [1, 4]. In the frequency domain, these factors are multiplied by the density and compressibility of the fluid.

To differentiate between high and low frequency regimes [1–3], the viscous and thermal layer thicknesses $\delta = \sqrt{2\eta/\omega\rho}$ and $\delta' = \delta/\sqrt{P_r}$ are compared, at a given frequency, with the effective radius of the pores r (ρ is the density of the saturation fluid, ω the pulsation frequency, P_r the Prandtl number, and η the viscosity of the fluid). The low frequency range is defined when the viscous [3] and thermal [1] skin thicknesses are great relatively to the pore radius. Otherwise, it is the high-frequency range. In the Darcy regime (flow without inertial effect), corresponding to the very low frequencies [11, 15], the air flow resistivity is the most important parameter describing the viscous losses caused by fluid/structure exchanges. In this case, the dynamic tortuosity α (ω) and the dynamic compressibility β (ω) are given by [2, 3, 13–15, 19]:

$$\alpha(\omega) = \frac{\sigma\phi}{j\omega\rho} \quad (1)$$

$$\beta(\omega) = \gamma \quad (2)$$

In these equations, $j^2 = -1$, ϕ is the porosity, σ is the flow resistivity, ρ is saturating fluid density and γ is the adiabatic constant.

Let us consider an acoustic wave arriving under normal incidence and striking a homogeneous porous material that occupies the region $0 \leq x \leq L$ (**Figure 1**). This wave generates an acoustic pressure field p and an acoustic velocity field v within the material that satisfies the following macroscopic equivalent fluid equations (along the x-axis):

$$\rho\alpha(\omega)j\omega v = \frac{\partial p}{\partial x}, \frac{\beta(\omega)}{K_a}j\omega p = \frac{\partial v}{\partial x} \quad (3)$$

where K_a is the compressibility modulus of the fluid.

The expression of a pressure field incident plane, unit amplitude, arriving at normal incidence to the porous material is given by

$$p^i(x, \omega) = e^{-j(kx-\omega t)}, \quad (4)$$

where $k = \frac{\omega}{c_0} = \omega\sqrt{\frac{\rho_0}{K_a}}$, k and c_0 are, respectively, the wave number and the wave velocity of the free fluid.

In the medium (1) (x < 0), the movement's results from the superposition of incident and reflected waves,

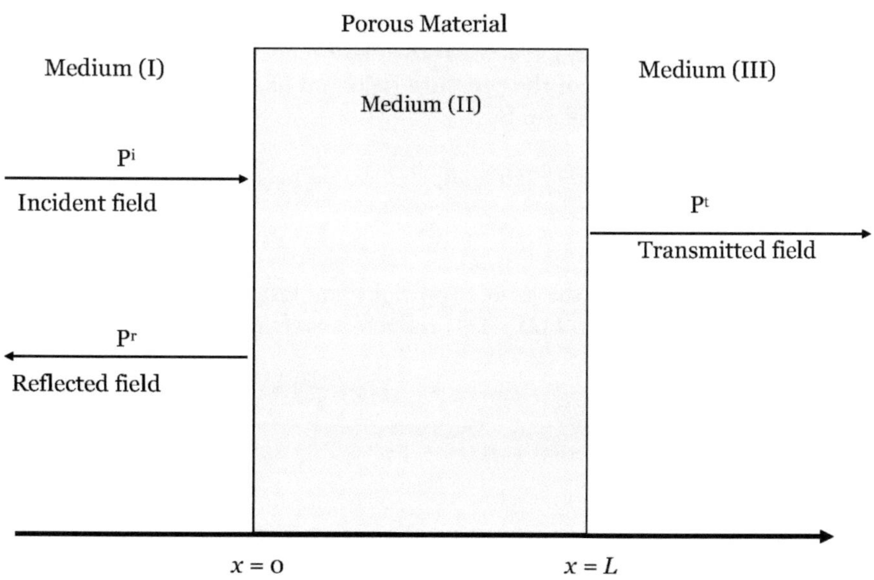

Figure 1.
Problem geometry.

$$p_1(x,\omega) = e^{-j(kx-\omega t)} + \tilde{R}e^{-j(-kx-\omega t)} \tag{5}$$

where $\tilde{R}$ is the reflection coefficient.

According to Eq. (3), the expression of the velocity field in the medium (1) is written:

$$v_1(x,\omega) = \frac{1}{Z_0}\left(e^{-j(kx-\omega t)} - \tilde{R}e^{-j(-kx-\omega t)}\right) \tag{6}$$

where $Z_0 = \sqrt{\rho_0 K_a}$ is the characteristic impedance of air.

In the medium (2) corresponding to the porous material, the expressions of the pressure and velocity field are:

$$p_2(x,\omega) = \tilde{A}e^{-j(\tilde{k}x-\omega t)} + \tilde{B}e^{-j(-\tilde{k}x-\omega t)} \tag{7}$$

$$v_2(x,\omega) = \frac{1}{\tilde{Z}_c}\left(\tilde{A}e^{-j(\tilde{k}x-\omega t)} - \tilde{B}e^{-j(-\tilde{k}x-\omega t)}\right) \tag{8}$$

In these expressions $\tilde{A}$ and $\tilde{B}$ are amplitude constants of the right-going and left-going waves, $\tilde{Z}_c$ and $\tilde{k}$ are the characteristic impedance and wave number, respectively, of the acoustic wave in the porous medium. These are two complex quantities:

$$\tilde{k} = \omega\sqrt{\frac{\tilde{\rho}}{\tilde{K}}} = \omega\sqrt{\frac{\rho_0\alpha(\omega)\beta(\omega)}{K_a}}, \text{and } \tilde{Z}_c = \sqrt{\tilde{\rho}\tilde{K}} = \sqrt{\frac{\rho_0 K_a \alpha(\omega)}{\beta(\omega)}} \tag{9}$$

Finally, in the medium (3), the expressions of the pressure and velocity fields of the wave transmitted through the porous material are,

$$p_3(x,\omega) = \tilde{T}e^{-j(k(x-L)-\omega t)}, \tag{10}$$

$$v_3(x,\omega) = \frac{1}{Z_0}\tilde{T}e^{-j(k(x-L)-\omega t)} \tag{11}$$

In these Eqs. ((10) and (11)) $\tilde{T}$ is the transmission coefficient.

The continuity conditions of the pressure field and of the velocity field at the boundary of the medium are given by:

$$p_1(0^-,\omega) = p_2(0^+,\omega)p_2(L^-,\omega) = p_3(L^+,\omega) \tag{12}$$

$$v_1(0^-,\omega) = \phi v_2(0^+,\omega)\phi v_2(L^-,\omega) = v_3(L^+,\omega) \tag{13}$$

the $\pm$ superscript denotes the limit from right and left, respectively. Using boundary and initial condition (12)–(13), reflected coefficient can be derived:

$$\tilde{R}(\omega) = \frac{\left(\phi^2 - \tilde{Z}^2\right) sinh\left(j\tilde{k}L\right)}{2\phi\tilde{Z}\ cosh\left(j\tilde{k}L\right) + \left(\phi^2 + \tilde{Z}^2\right) sinh\left(j\tilde{k}L\right)} \tag{14}$$

where $\tilde{Z} = \frac{\tilde{Z}_c}{Z_0} = \sqrt{\frac{\alpha(\omega)}{\beta(\omega)}}$ is the normalized characteristic impedance of the material.

Using the expressions of the dynamic tortuosity $\alpha(\omega)$ and the dynamic compressibility $\beta(\omega)$ given by Eq.(1), the expression (14) of the reflection coefficient becomes:

$$R(\omega) = \frac{(1 - C_1^2\omega)\, sinh\, (LC_2\sqrt{j\omega})}{2C_1\sqrt{j\omega}\ cosh\, (LC_2\sqrt{j\omega}) + (1 + C_1^2\omega)\, sinh\, (LC_2\sqrt{j\omega})} \tag{15}$$

where

$$C_1 = \sqrt{\frac{\gamma\rho\phi}{\sigma}}\ \text{et}\ C_1 = \sqrt{\frac{\gamma\sigma\phi}{K_a}} \tag{16}$$

By doing the Taylor series expansion of the reflection coefficient (Eq. (15)), limited to the first approximation, the reflection coefficient expression is written at very low frequencies (see appendix):

$$R = \frac{1}{1 + \frac{2}{L\sigma}\sqrt{\rho K_a}} \tag{17}$$

This simplified expression of the reflection coefficient is independent of the frequency and porosity of the material, and depends only on the flow resistivity σ and the thickness L of the material.

The incident p^i and reflected p^r fields are related in the frequency domain by the reflection coefficient R:

$$p^r_{sim}(x, \omega) = R\, p^i(x, \omega) \tag{18}$$

The time-domain simulated reflected signals $\mathcal{P}^r_{sim}(x, t)$ are obtained numerically by taking the inverse Fourier transform $\mathcal{F}^{-1}$ of (18),

$$\mathcal{P}^r_{sim}(x, t) = \mathcal{F}^{-1}(R\, p^i(x, \omega)) \tag{19}$$

3. Inverse problem

The simplified expression of the reflection coefficient obtained at low frequency (Eq.(17)) depends only on the flow resistivity σ and thickness L of the medium. Our objective is to find this two parameters simultaneously, supposedly unknown, by minimizing between the simulated reflected signal given by the expression (18) and the experimental reflected signal. The inverse problem then consists in finding the flow resistivity σ and thickness L of porous samples that minimize the function:

$$U(\sigma, L) = \sum_{i=1}^{i=N} \left(p^r_{exp}(\omega) - p^r_{sim}(\sigma_i, L_i, \omega)\right)^2 \tag{20}$$

Wherein $p^r_{sim}(\sigma_i, L_i, \omega)$ are the discrete sets values of the simulated reflected signal and $p^r_{exp}(x, \omega)$ are the discrete sets of values of the experimental reflected signal. The minimization is made in frequency domain. The experimental setup [15] is shown in **Figure 2**. The tube length is adaptable to avoid reflection, and to permit the propagation of transient signals, according to the desired frequency range. For measurements in the frequency range (20–100) Hz, a length of 50 m is sufficient. The tube diameter is 5 cm (the cut-off of the tube $f_c \sim 4$ kHz). A sound source Driver unit "Brand" constituted by loudspeaker Realistic 40–9000 is used. Tone-bursts are provided by Standard Research Systems Model DS345–30 MHz synthesized function generator. The signals are amplified and filtered using model

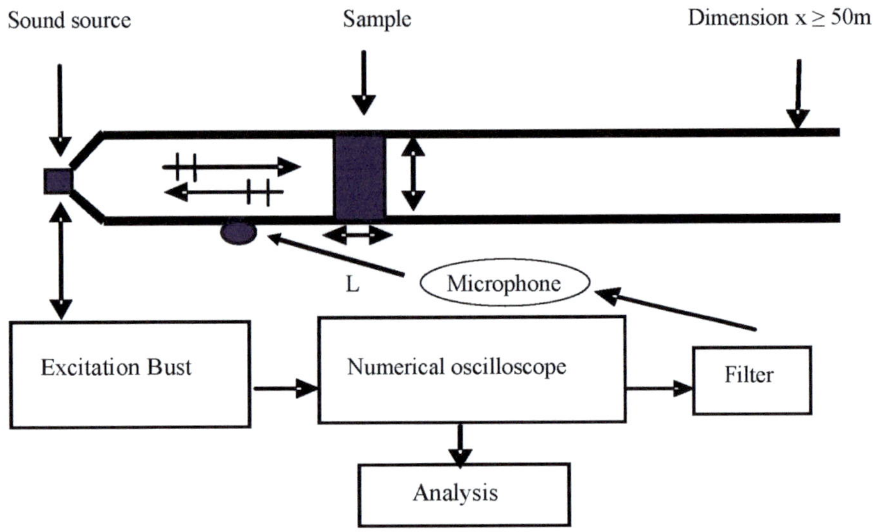

Figure 2.
The experimental set up.

SR 650-Dual channel filter, Standford Research Systems. The signals (incident and reflected) are measured using the same microphone. The incident signal is measured by putting a total reflector [15] in the same position as the porous sample. **Figures 3**, **4** show the incident and reflected signals and their spectrum of the two samples in frequency bandwidth of 50 Hz.

The inverse problem is solved for two cylindrical polyurethane (PU) foams named (M1) and (M2) with a rigid frame and an open cell structure. Polyurethane foam is a leading member of the large and very diverse family of polymers or plastics and has many uses in the automotive sector and for the thermal insulation of buildings. The flow resistivity and thicknesses of the two samples M1 and M2 are measured by conventional methods [20, 21] and given in **Table 1**.

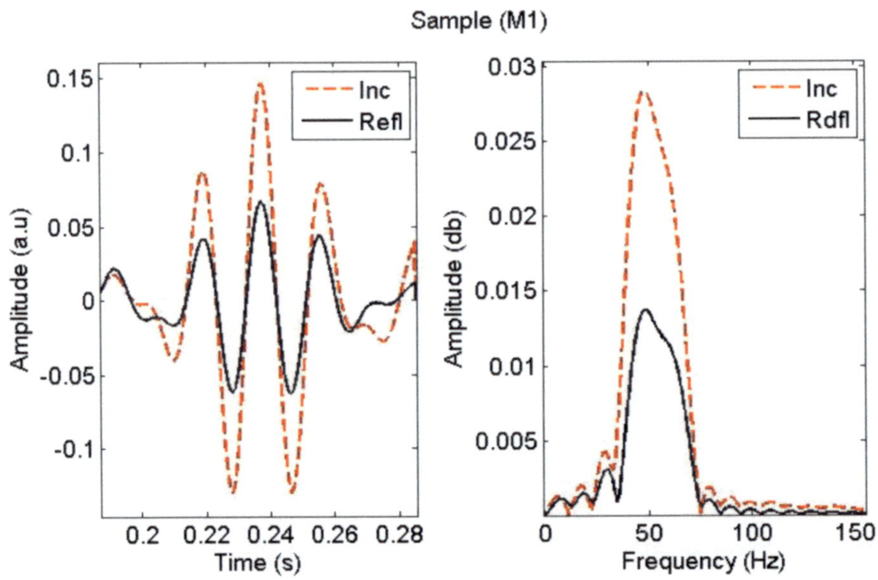

Figure 3.
The incident and reflected signals and their spectrum of the sample (M1) in frequency bandwidth of 50 Hz.

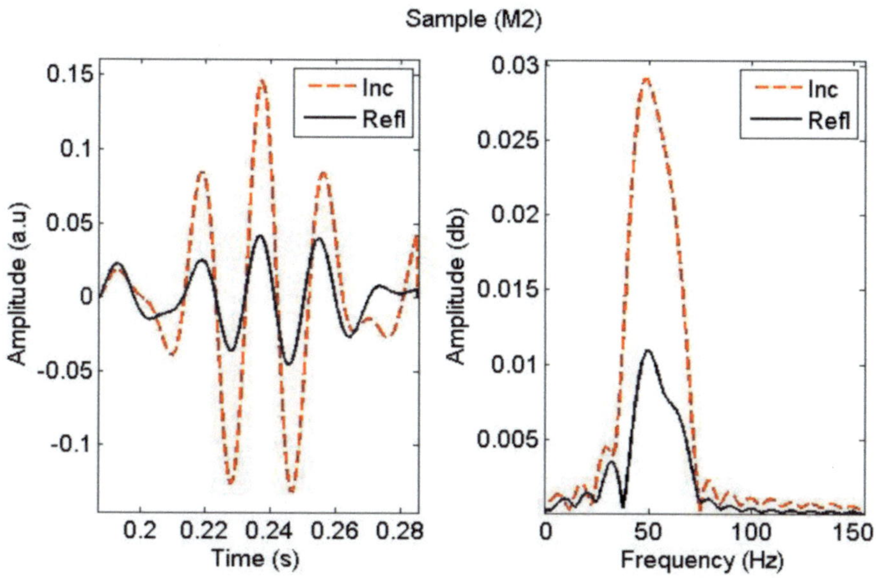

Figure 4.
The incident and reflected signals and their spectrum of the sample (M2) in frequency bandwidth of 50 Hz.

Samples	M1	M2
Thickness (cm)	2.6 ± 0.5	5.0 ± 0.5
Resistivity ($Nm^{-4}s$)	$27{,}500 \pm 500$	7500 ± 500

Table 1.
Flow resistivity and thickness of the sample M1 and M2.

The inverse problem is to find the parametric vector $V = \{\sigma, L\}$ which satisfies the conditions:

$$\begin{cases} U(\sigma, L) \to 0 \\ LV \leq V \leq UV \end{cases} \tag{21}$$

where LV and UV are the lower and upper bounds that limit the research domain on the adjustable parametric vector V. For plastic foam samples, the value of the flow resistivity is greater than 3000 $Nm^{-4}s$. The lower and upper limits in Eq. (20) can be built from the following constraints:

$$\begin{cases} \sigma \geq 3000\ Nm^{-4}s \\ 0 \leq L \leq 10\ cm \end{cases} \tag{22}$$

The inverse problem is solved by the last-square method. For its iterative solution, we used the simplex search method [22–26] which does not require numerical or analytic gradient. The flow resistivity and the thickness are inverted using experimental reflected signals by two PU porous material samples (M1 and M2). The variations in the cost function present one clear minimum corresponding to the solution of the inverse problem. **Figures 5, 6** show the variation of the cost function U when varying the flow resistivity and the thickness in different frequency bandwidths for the samples (M1, M2). The results of the inverse problem

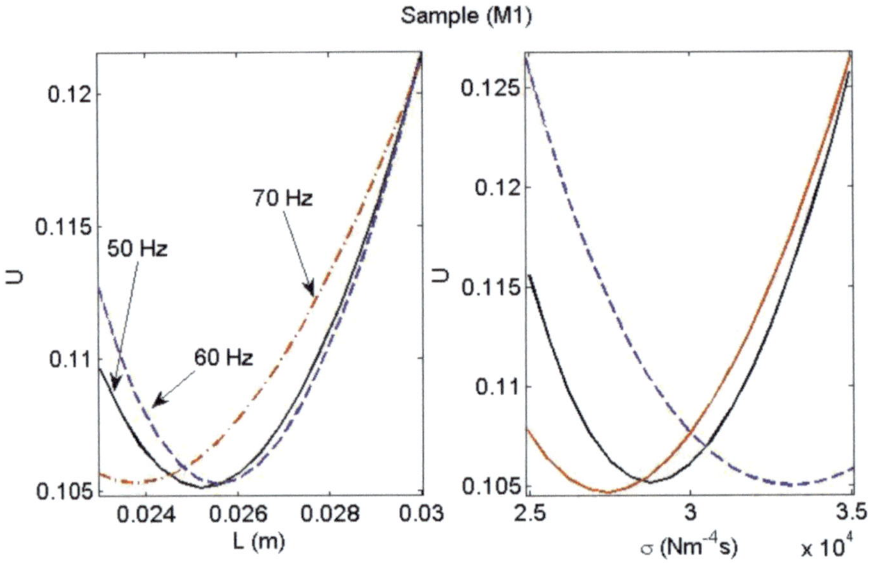

Figure 5.
Variation of the cost function U when varying the flow resistivity and the thickness in different frequency bandwidths for the samples M1.

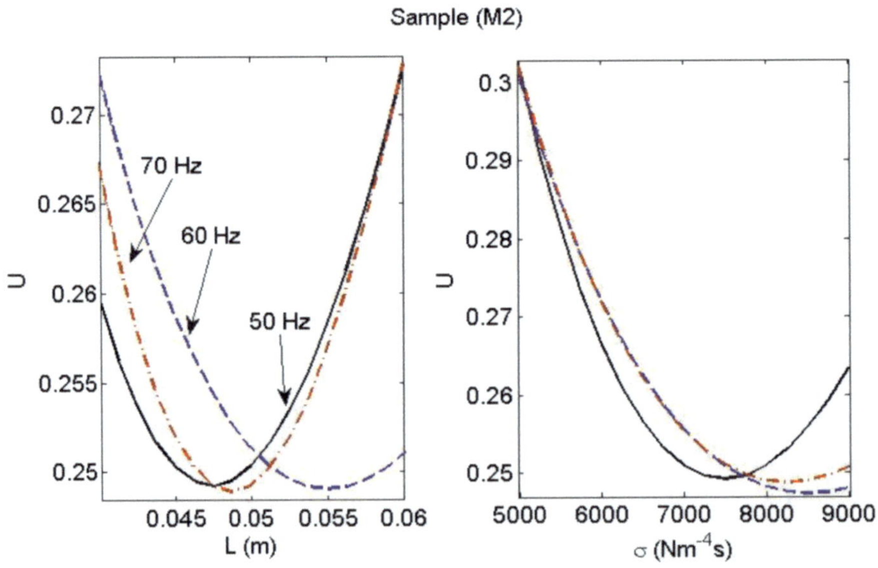

Figure 6.
Variation of the cost function U when varying the flow resistivity and the thickness in different frequency bandwidths for the samples M2.

are summarized in **Table 2**, in which inverted values of flow resistivity and thickness are given for different frequency bandwidths. A comparison between an experimental reflected signal and simulated reflected signal is given in **Figures** 7, 8 for the optimized values of the inverted flow resistivity and thickness of the porous samples (M1, M2), respectively. The frequency bandwidth of the incident signals is (40–60) Hz. It can be seen that the agreement between experiment and theory is good for the two samples and the inverted values are close to those given by conventional methods.

Samples	*Frequency (Hz)*	*Thickness (cm)*	*Resistivity* ($Nm^{-4}s$)
M1	50	2.52	28,750
	60	2.56	33,125
	70	2.39	27,500
M2	50	4.75	7500
	60	5.55	8500
	70	4.88	8250

Table 2.
Inverted parameters obtained of the flow resistivity and the thickness of the two samples M1 and M2.

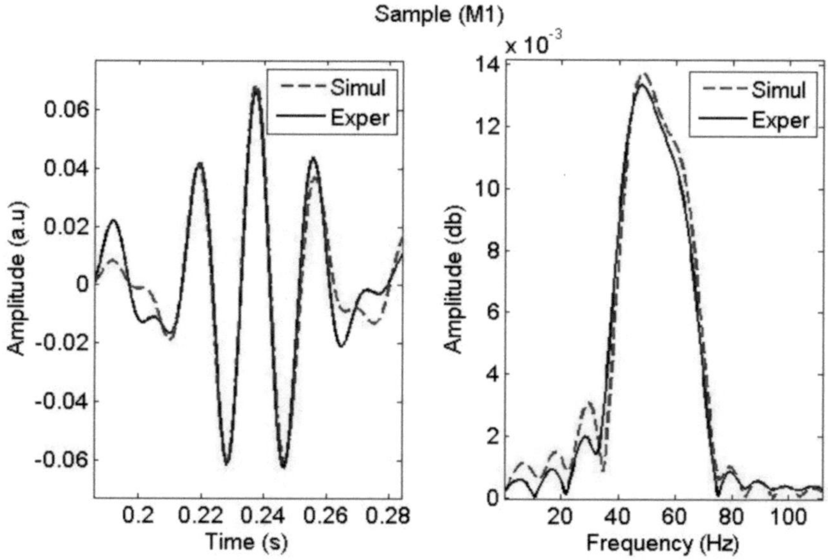

Figure 7.
Comparison between an experimental reflected signal and simulated reflected signal of the sample M1.

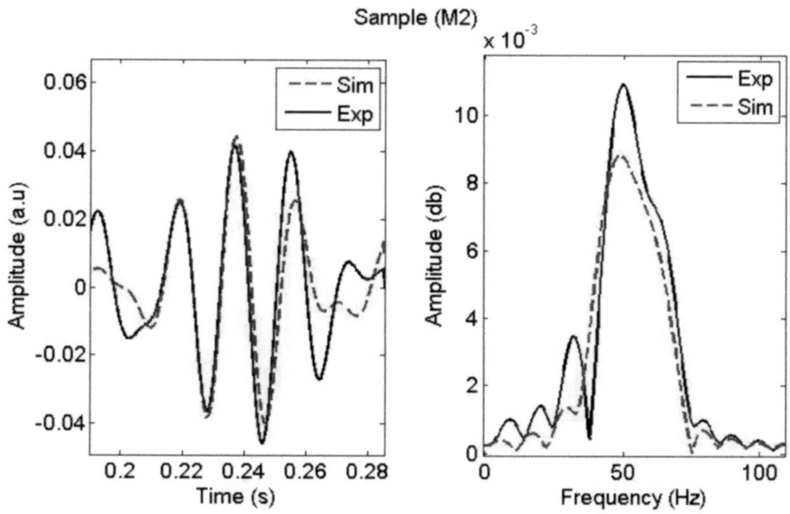

Figure 8.
Comparison between an experimental reflected signal and simulated reflected signal of the sample M2.

4. Conclusion

Simultaneous determination of the flow resistivity and the thickness of a rigid porous medium are obtained by solving the inverse problem using experimental signals at very low frequencies. The model is based on a simplified expression of the reflection coefficient which is independent on frequency and porosity and depends only on the flow resistivity and thickness of the medium. Two plastic foam samples having different values of flow resistivity and different thickness are tested using this proposed method. The results are satisfactory and the inverted values of flow resistivity and thickness are close to those given by conventional methods. The advantage of the proposed method is that the two parameters, resistivity and thickness of the porous medium, were determined simultaneously without knowing previously any other parameter describing the porous medium, including its porosity. The suggested method opens new perspectives for the acoustic characterization of porous materials.

Acknowledgements

This work is funded by the university training research project (PRFU) under number: B00L02UN440120200001 and by the General Direction of Scientific Research and Technological Development (DGRSDT).

Conflict of interest

The authors declare that they have no conflict of interest.

Appendix: Taylor series expansion of the reflection coefficient

The reflection coefficient given by Eq. (15) can be rewritten as [14, 15]:

$$R(\omega) = \left(\frac{1 - C_1^2 j\omega}{1 + C_1^2 j\omega}\right)\left(\frac{1}{1 + \coth\left(LC_2\sqrt{j\omega}\right)\frac{2C_1\sqrt{j\omega}}{1+C_1^2 j\omega}}\right) \tag{A.1}$$

where C_1 and c_2 are given by Eq. (16). Taylor's limited serial expansion in the vicinity of zero of the expressions $\left(\frac{1-C_1^2 j\omega}{1+C_1^2 j\omega}\right)$, $\frac{2C_1\sqrt{j\omega}}{1+C_1^2 j\omega}$ and $\coth\left(LC_2\sqrt{j\omega}\right)$ is given by:

$$\left(\frac{1 - C_1^2 j\omega}{1 + C_1^2 j\omega}\right) = 1 - 2C_1^2 j\omega + \mathcal{O}\left((j\omega)^2\right), \tag{A.2}$$

$$\frac{2C_1\sqrt{j\omega}}{1 + C_1^2 j\omega} = 2C_1\sqrt{j\omega} - 2C_1^3 (j\omega)^{3/2} + \mathcal{O}\left((j\omega)^{5/2}\right), \tag{A.3}$$

and,

$$\coth\left(LC_2\sqrt{j\omega}\right) = \frac{1}{LC_2\sqrt{j\omega}} + \frac{1}{3}LC_2 j\omega - \frac{1}{45}L^3C_2^3(j\omega)^{3/2} + \mathcal{O}\left((j\omega)^{5/2}\right), \tag{A.4}$$

with,

$$\coth\left(LC_2\sqrt{j\omega}\right)\frac{2C_1\sqrt{j\omega}}{1+C_1^2 j\omega}=\frac{2C_1}{LC_2}+\frac{2}{3}C_1C_2L\left(1-\frac{3C_1^2}{L^2C_2^2}\right)j\omega+\mathcal{O}\left((j\omega)^2\right), \tag{A.5}$$

and,

$$\frac{1}{1+\coth\left(LC_2\sqrt{j\omega}\right)\frac{2C_1\sqrt{j\omega}}{1+C_1^2 j\omega}}=\frac{1}{1+\frac{2C_1}{LC_2}}-\frac{\frac{2}{3}L^3C_2^3C_1\left(1-\frac{3C_1^2}{L^2C_2^2}\right)}{\left(1+\frac{2C_1}{LC_2}\right)^2}j\omega+\mathcal{O}\left((j\omega)^2\right), \tag{A.6}$$

Using Eqs. (A.1), (A.2) and (A.6), one obtains

$$R(\omega)=\left(\frac{1}{1+\frac{2C_1}{LC_2}}\right)\left(1-\frac{\frac{2}{3}LC_1C_2\left(1+\frac{3C_1}{LC_2}+3\frac{3C_1^2}{L^2C_2^2}\right)}{\left(1+\frac{2C_1}{LC_2}\right)}j\omega+\mathcal{O}\left((j\omega)^2\right)\right) \tag{A.7}$$

As a first approximation, at very low frequencies, the reflection coefficient (A.7) is given by the first term

$$R=\frac{1}{1+\frac{2C_1}{LC_2}}=\frac{1}{1+\frac{2}{L\sigma}\sqrt{\rho K_a}} \tag{A.8}$$

Author details

Mustapha Sadouki
Acoustics and Civil Engineering Laboratory, Matter Sciences Department, Faculty of Sciences and Technology, Khemis-Miliana University, BP. 44225 Ain Defla, Algeria

*Address all correspondence to: mustapha.sadouki@univ-dbkm.dz

References

[1] Allard. J. F, Propagation of Sound in Porous Media Modelling. Sound Absorbing Materials. Elsevier, London, UK, 1993, p. 1–284; DOI 10.1007/ 978-94-011-1866-8

[2] Lafarge D, Materials and Acoustics Handbook, edited by M. Bruneau and C. Potel. ISTE-Wiley, London, 2009, p. 149–202.

[3] Johnson D. L, Koplik J, and Dashen R, "Theory of dynamic permeability and tortuosity in fluid-saturated porous media," J. Fluids Mech, 1987; 176–379; https://doi.org/10.1017/ S0022112087000727

[4] Champoux Y, and Allard J. F, Dynamic tortuosity and bulk modulus in air-saturated porous media. J. Appl. Phys.1991; 70: 1975–1979; https://doi. org/10.1063/1.349482

[5] Zwikker C, and Kosten C. W, Sound Absorbing Materials. Elsevier, 1949, New York.

[6] Henry M, Lemarinier P, Allard J. F, Bonardet J. L, and Gedeon A, Evaluation of the characteristic dimension for porous sound absorbing materials. J. Appl. Phys, 1995;77: 17–20.

[7] Tizianel J, Allard J. F, Castagnéde B, Ayrault C, Henry M and Moussatov A, Transport parameters and sound propagation in an air-saturated sand. J. Appl. Phy. Am.1999; 86: 5829; https:// doi.org/10.1063/1.371599

[8] Attenborough K, On the acoustic slow wave in air-filled granular media. J. Acoust. Soc. Am. 1987; 81, 93; https:// doi.org/10.1121/1.394938

[9] Nagy P. B, Adler L, and Bonner B. P, Slow wave propagation in air-filled porous materials and natural rocks. Appl. Phys. Lett. 1990;56: 2504; https:// doi.org/10.1063/1.102872

[10] Delaney M, and Bazley E, Acoustical properties of fibrous absorbent materials, Appl. Acoust. 1970;3: 105–116; https://doi.org/10.1016/0003-682X (70)90031-9

[11] Sadouki M, Fellah Z. E. A, Berbiche A, Fellah M, Mitri F. G, Ogam E, and Depollier C, Measuring static viscous permeability of porous absorbing materials, J. Acoust. Soc. Am, 2014; 135: 3163; https://doi.org/10.1121/1.4874600

[12] Sadouki M, Experimental characterization of rigid porous material via the first ultrasonic reflected waves at oblique incidence. J.ApAcoust.2018;133: 64–72; https://doi.org/10.1016/j.apac oust.2017.12.010

[13] Sadouki M, Experimental Measurement of the porosity and the viscous tortuosity of rigid porous material in low frequency. Journal of Low Frequency Noise Vibration and Active Control, 2018;37(2): 385–393; https://doi.org/10.1177% 2F1461348418756016

[14] Sadouki M, Theoretical modeling of acoustic propagation in an inhomogeneous porous medium. (in French),(Ph. D. Thesis), Université des sciences et de la technologie ' Houari Boumediène', Algiers, Algeria; 2014,

[15] Berbiche A, Sadouki M, and Fellah Z.E.A, Fellah M, Mitri F.G, Ogam E, Depollier C, Experimental determination of the viscous flow permeability of porous materials by measuring reflected low frequency acoustic waves. J. Appl. Phys. 2016; 119: 014906; https://doi.org/10.1063/ 1.4939073

[16] Biot M. A, The theory of propagation of elastic waves in a fluid-saturated porous solid, low frequency range. J. Acoust. Soc. Am, 1956; 28:168; https://doi.org/10.1121/1.1908241

[17] Sadouki M, Fellah M, Fellah Z.E.A, and Ogam E, Depollier C. Ultrasonic propagation of reflected waves in cancellous bone: Application of Biot theory. ESUCB 2015, 6th European Symposium on Ultrasonic Characterization of Bone, 10–12 June 2015; Corfu, (Greece). p.1–4; DOI: 10.1109/ESUCB.2015.7169900

[18] Sadouki M, Experimental characterization of human cancellous bone via the first ultrasonic reflected wave – Application of Biot's theory, J. Ap.Acoust, 2020; 107237: 163; https://doi.org/10.1016/j.apacoust.2020.107237

[19] Sadouki M, Direct problem for reflected wave at the first interface of a rigid porous medium in Darcy's regime. Proc. Mtgs. Acoust. 2015; **25**: 045004; https://doi.org/10.1121/2.0000200

[20] Leonard R. W, Simplified flow resistance measurements. J. Acoust. Soc. Am.,1946; 17: 240; https://doi.org/10.1063/1.2099510

[21] Stinson M. R and Daigle G. A, Electronic system for the measurement of flow resistance, J. Acoust. Soc. Am. 1988;83: 2422. https://doi.org/10.1121/1.396321

[22] Lagarias JC, Reeds JA, Wright MH, et al. Convergence properties of the Nelder-Mead Simplex method in low dimensions. SIAM J Optim 1998; 9: 112–147; https://doi.org/10.1137/S1052623496303470

[23] Sadouk M, Thickness measurement of rigid porous material through reflected acoustic waves at Darcy's regime, Proc. Mtgs. Acoust. 2018;**35:** 045003; https://doi.org/10.1121/2.0000973.

[24] Sadouki M., Experimental measurement of tortuosity, viscous and thermal characteristic lengths of rigid porous material via ultrasonic transmitted waves, Proc. Mtgs. Acoust. 2018; **35:** 045005; https://doi.org/10.1121/2.0000991.

[25] Sadouki M, Berbiche A, Fellah M, Fellah Z E. A and Depollier C. Characterization of rigid porous medium via ultrasonic reflected waves at oblique incidence, Proc. Mtgs. Acoust. 2015; **25:** 045005; https://doi.org/10.1121/2.0000201

[26] Sadouki M, Berbiche A, Fellah M, Fellah Z E. A and Depollier C, Measurement of tortuosity and viscous characteristic length of double-layered porous absorbing materials with rigid-frames via transmitted ultrasonic-wave, Proc. Mtgs. Acoust. 2015; **25**: 045001; https://doi.org/10.1121/2.0000184.